REGULAR MAPPINGS AND THE SPACE OF HOMEOMORPHISMS
ON A 3-MANIFOLD

by

Mary-Elizabeth Hamstrom

1. Introduction. In our paper [5] Eldon Dyer and I introduced the notion of a
completely regular mapping f of a metric space X onto a metric space Y (for definitions,
see section 2) and were able to prove, with certain added hypotheses on X, Y, and the inverses under f, that (X, f, Y) is a locally trivial fibre space. Under some conditions, f is
the projection mapping of a direct product. In [5] and [6] it was shown that if f is a
0-regular mapping of a metric space X onto a metric space Y and M is a compact 2-
manifold with boundary to which each inverse under f is homeomorphic, then f is completely regular. In [7] I proved that if f is homotopy 2-regular and M is a compact 3-
manifold with boundary which is imbeddable in E^3, then f is completely regular. Thus
it may be shown that if M is a 2-manifold or a 3-cell, X is complete and Y has finite
covering dimension, then (X, f, Y) is a locally trivial fibre space and if Y is locally compact, separable and contractible, then X is homeomorphic to the direct product Y × M,
f corresponding to the projection map of Y × M onto Y. If M is a 3-manifold other than
a 3-cell and is imbeddable in E^3, the restriction that Y be one-dimensional had to be
made. In the present paper, this restriction is removed and these theorems are extended to the case where M is any compact 3-manifold with boundary in which every
homotopy 3-cell is a 3-cell. As a tool in proving this and as a result of interest in itself it is proved, using the techniques of [5], [6], and [7], that the space of homeomorphisms of a compact 3-manifold with boundary onto itself is LC^n for each n (see section
5). This duplicates, in part, results obtained independently and previously by J. M.
Kister and G. M. Fisher [4] and [8]. See also Sanderson's paper [15].

2. Some results of an earlier paper and some definitions. The reader is referred
to [7], which contains special cases of many of the theorems in this paper. Since some

Presented to the American Mathematical Society, January 20, 1960. This research was
supported by National Science Foundation Grant NSF G-8202.

of the proofs in [7] may be used here without essential change, they will not be repeated. Two lemmas from [7] (it is not pretended that they are new results) are stated here for convenient reference. The numbers from [7] are retained, an asterisk added to prevent confusion with the numbering here.

Lemma 2.4*. If M is a compact 2-manifold with boundary and ϵ is a positive number, then there is a positive number δ such that each mapping of M into itself that moves no point as much as δ is ϵ-homotopic in M to the identity map.

Lemma 2.5*. If f is a mapping of a compact 2-manifold M onto a compact 2-manifold N, f is not homotopic in N to a map carrying M onto a proper subset of N, and A is an annulus in N, then some simple closed curve in $f^{-1}(A)$ is mapped essentially into A (i.e. f, restricted to this curve, is not homotopic to 0 in A).

In this paper, the term n-manifold denotes a separable, metric (not necessarily connected) space M each point of which is contained in an open set which is homeomorphic to E^n, Euclidean n-space. A separable metric space M is an n-manifold with boundary if each of its points is contained in an open set whose closure is homeomorphic to a (closed) n-cell. The set of points of M which are not contained in open sets homeomorphic to E^n is called the boundary of M and is denoted by bdry M. The set of non-boundary points of M will be denoted by int M. The n-cell in E^n consisting of those points $(x_1,\dots,x_n)$ such that $0 \leq x_i \leq 1$ for $i = 1,\dots,$ n will be denoted by R^n and its boundary will be denoted by S^{n-1}. The set R^1 will usually be denoted by I.

A sequence whose elements are x_1, $x_2,\dots$ will be denoted by the symbol $\{x_i\}_i$ or, if no ambiguity results, by $\{x_i\}$.

Definition 2.1. A proper mapping f of a metric space X onto a metric space Y is homotopy n-regular (h-n-regular) provided that it is true that f is open and if x is a point of X and ϵ is a positive number, then there is a positive number δ such that each mapping of a k-sphere, $k \leq n$, into $S(x,\epsilon) \cap f^{-1}(y)$, $y \in Y$, is homotopic to 0 in $S(x,\epsilon) \cap f^{-1}(y)$. (A proper map is one for which inverses of compact sets are compact. The symbol $S(x,\delta)$ denotes the set of points with distance from x less than δ.)

Definition 2.2. The mapping f is said to be completely regular provided that if $\epsilon > 0$ and $y \in Y$, then there is a positive number δ such that $d(y,y') < \delta$, $y' \in Y$, implies that there is a homeomorphism of $f^{-1}(y)$ onto $f^{-1}(y')$ which moves no point as much as ϵ

(i.e. an ϵ-homeomorphism).

Definition 2.3. If f is an h-n-regular (completely regular) mapping of a compact space X onto a space Y and Y consists of the points $y_0, y_1, \ldots$ of a sequence which converges to y_0, then the sequence $\{f^{-1}(y_i)\}$ is said to <u>converge h-n-regularly</u> (<u>completely regularly</u>) to $f^{-1}(y_0)$.

In sections 2,3, and 4 of this paper, M_0, $M_1, \ldots$ will denote the elements of a sequence of mutually exclusive compact connected 3-manifolds with boundary converging h-2-regularly to M_0, the union of whose elements is a compact metric space. It will be assumed that the boundaries, K_0, $K_1, \ldots$ of these 3-manifolds are mutually homeomorphic. The space $\bigcup M_i$ may be imbedded in some Euclidean space E^n in such a way that M_0 is a polyhedral subset of E^n. See [9], 3.3. It will follow readily from the proofs that the assumption of connectivity of the M_i need not be made, but the proofs are somewhat simpler if this assumption is made. The 3-manifold M_i may be triangulated by means of a triangulation Γ_i (see [12]). In each M_i, polyhedra, piecewise linear mappings, et cetera will be defined relative to Γ_i and its refinements. No relationships among the Γ_i will be assumed. Distances will be ordinary distances in E^n. <u>Unless it is explicitly stated otherwise, any subset of M_i that is referred to will be polyhedral.</u>

Much use will be made of the following lemma from [7], which, although stated there for manifolds imbeddable in E^3, does not require this restriction in its proof.

Lemma 2.6*. If ϵ is a positive number, then there is an integer N such that if $i > N$, then there is a piecewise linear ϵ-mapping of M_0 into int M_i and there is a piecewise linear ϵ-mapping of M_i into int M_0.

Definition 2.4. A <u>regular neighborhood in M_i</u> of a polyhedron P_i is a closed neighborhood of which P_i is a (strong) deformation retract and whose boundary is a compact polyhedral 2-manifold (with boundary if P_i intersects K_i). A <u>regular neighborhood in E^n</u> of a polyhedron P in M_0 is a closed neighborhood U such that P and $U \cap M_0$ are deformation retracts of U and $U \cap M_0$ is a regular neighborhood of P in M_0. The mappings of U onto P and $U \cap M_0$ resulting from the deformation will be called the <u>natural</u> or <u>projection mappings</u> of U onto P and $U \cap M_0$. For each positive number ϵ there exist such neighborhoods U for which no point is moved as much as ϵ during the deformation of U onto $U \cap M_0$, U onto P or $U \cap M_0$ onto P. Such neighborhoods will be called <u>regular ϵ-neighborhoods</u>. Their existence was demonstrated by J. H. C. Whitehead in [19].

<u>Definition 2.5</u>. If J is a subpolyhedron of P above, U' is a regular neighborhood of J in E^n and r and q are the deformations of U, as above, into $U \cap M_0$ and $U \cap M_0$ into P, then U' is said to be <u>consistently imbedded</u> in U provided that (1) $U' \cap M_0$ and $(U - U') \cap M_0$ are deformation retracts of U' and U - U' under r. (2) $U' \cap P$ and $(U - U' \cap P$ are deformation retracts of $U' \cap M_0$ and $(U - U') \cap M_0$ under q, and (3) J is a deformation retract of $U' \cap P$.

<u>Definition 2.6</u>. If H_0 is a compact polyhedral p-manifold, p = 1,2, in M_0 and $\{H_i\}$ is a sequence of compact polyhedral p-manifolds converging to H_0 such that for each i, H_i lies in M_i, then the sequence $\{H_i\}$ is said to <u>converge strongly</u> to H_0 if it is true that for each regular neighborhood in E^n, V_0, of H_0 and for sufficiently large i, the non-trivial p-cycle, mod 2, carried by H_i fails to bound in $V_0 \cap M_i$. If $\{D_i\}$ is a sequence of discs (annuli) converging to the disc (annulus) D_0 in M_0 and for each i, $D_i \subset M_i$, then the sequence $\{D_i\}$ is said to <u>converge strongly</u> to D_0 if the sequence $\{$bdry $D_i\}$ converges strongly to bdry D_0. (This definition varies slightly from that in [7], which uses cycles over the integers. Since non-orientability arises here, mod 2 homology is used.)

<u>Definition 2.7</u>. If $\{D_i\}$ is a sequence of spheres (tori, simple closed curves, discs, annuli) converging strongly to the sphere (torus, simple closed curve, disc, annulus) D_0 and for each i, $D_i \subset M_i$, then $\{D_i\}$ is said to <u>converge properly</u> to D_0 if for each regular neighborhood in E^n, V_0, of D_0 and sufficiently large i, it is true that the natural map of V_0 onto D_0, restricted to D_i, is homotopic in D_0 to a homeomorphism onto D_0, the image of bdry D_i remaining in a regular neighborhood of bdry D_0 under this homotopy.

3. <u>Completely regular convergence of</u> $\underline{K}_i$ <u>to</u> $\underline{K}_0$.

<u>Theorem 3.1</u>. If C_0 is a component of K_0 then there is a sequence $\{C_i\}$ converging strongly to C_0 such that for each i, C_i is a component of K_i.

<u>Proof</u>. Denote by V_0 a regular neighborhood of C_0 in M_0 which is homeomorphic to the product $C_0 \times I$ and is bounded by C_0 and a compact 2-manifold C_0' lying in int M_0

and homeomorphic to C_0. That such a neighborhood exists follows from [11]. Let γ_0 and γ_0' denote the non-trivial 2-cycles mod 2 carried by C_0 and C_0'. Denote by W_0 a regular neighborhood of V_0 in E^n and by U_0 and U_0' mutually exclusive regular neighborhoods of C_0 and C_0' in E^n, each consistently imbedded in W_0. Let p, q, q' denote the natural projection mappings of W_0, U_0 and U_0' onto V_0, C_0 and C_0'. Finally, let $\{x_i\}$ be a sequence converging to a point x of int M_0 - $(W_0 \cap$ int $M_0)$ such that for each i, x_i is a 3-cell in Γ_i or a suitable refinement of Γ_i.

From Lemma 2.6* of [7] it follows that there are a sequence of positive numbers, $\{\epsilon_i\}$, converging to 0 and a sequence $\{g_i\}$ such that for each i, g_i is a piecewise linear ϵ_i-mapping of M_0 into int M_i. For sufficiently large i, $g_i(C_0) \subset U_0 \cap M_i$ and $g_i(C_0')$ $\subset U_0' \cap M_i$. Denote by U_i and U_i' regular neighborhoods of $g_i(C_0)$ and $g_i(C_0')$ in $U_0 \cap$ int M_i and $U_0' \cap$ int M_i so chosen that the sequences $\{U_i\}$ and $\{U_i'\}$ converge to C_0 and C_0'.

Suppose that A_i, carrying a non-trivial 2-cycle mod 2, γ_i, which bounds in $U_0' \cap M_i$, is a component of bdry U_i'. Then γ_i bounds in M_i - int x_i. Consequently A_i separates M_i into two components, the closure of one of which is a 3-manifold $T(A_i)$ which lies in M_i - x_i and is bounded by A_i. Since γ_i bounds mod 2 in $T(A_i)$, $T(A_i)$ does not intersect bdry M_i and, since γ_i bounds in $U_0' \cap$ int M_i, $T(A_i) \subset U_0' \cap$ int M_i.

If each component, A_i, of bdry U_i' carries a non-trivial 2-cycle which bounds mod 2 in $U_0' \cap M_i$, then for sufficiently large i, there is a component B_i of bdry U_i' such that the closure V_i' of the component of M_i - B_i not containing x_i contains $T(A_i)$ for each component A_i of bdry U_i'. Then V_i' contains U_i' and $g_i(C_0')$ and lies in $U_0' \cap M_i$.

Denote by U_i'' the union of U_i and the components, if any, of M_i - U_i which lie in $U_0 \cap M_i$. Then U_i'' contains $g_i(C_0)$ and, since the convergence of $\{M_i\}$ to M_0 is 0-regular, M_i - U_i'' is connected for sufficiently large i. Thus M_i - $(U_i'' \cup V_i')$ is connected. However, $g_i(\gamma_0)$ is homologous mod 2 to $g_i(\gamma_0')$ in $W_0 \cap M_i$ and in M_i - x_i. But, since the cell x_i has coefficient 0 in this homology, so does each cell in M_i - $(U_i'' \cup V_i')$. Thus $g_i(\gamma_0)$ is homologous to $g_i(\gamma_0')$ in $U_i'' \cup V_i'$. (This is a conse-quence of Lemma 2.6* of [7] cited above.) Since U_i'' and V_i' are mutually exclusive, $g_i(\gamma_0')$ bounds in V_i' and $g_i(\gamma_0)$ bounds in U_i''. Thus $q_i'g_i(\gamma_0')$ bounds in C_0', where q_i' is a piecewise linear ϵ_i- approximation to $q'|U_i'$. But $q_i'g_i|C_0'$ is, for sufficiently large i, homotopic to the identity mapping (Lemma 2.4* of [7]) and thus $q_i'g_i(\gamma_0')$ does not bound in C_0'. This contradiction implies that for sufficiently large i some component C_i' of bdry U_i' carries a non-trivial 2-cycle γ_i' which does not bound mod 2 in $U_0' \cap M_i$. Since U_0' was chosen arbitrarily, a sequence $\{C_i'\}$ is obtained converging to a subset of C_0'. If this subset is proper, then for sufficiently large i, $q_i'(C_i')$ is a proper subset of C_0' and $q_i'(\gamma_i')$ bounds mod 2 in C_0'. Consequently $g_i q_i'(\gamma_i')$ bounds in $U_0' \cap M_i$ and thus γ_i' bounds in $U_0' \cap M_i$ (Lemma 2.6* of [7]). This is a contradiction. Therefore $\{C_i'\}$ converges strongly to C_0'. Since this argument may be applied to each element of a sequence of regular neighborhoods of C_0 in M_0 converging to C_0, there may be obtained a sequence $\{C_i''\}$ converging strongly to C_0 such that for each i, C_i'' is a compact 2-manifold in int M_i.

Let V_0' be a regular neighborhood of V_0 in M_0 which is homeomorphic to $C_0 \times I$ and lies in W_0. Since $p_i(C_i' \cup C_i'') \subset V_0'$, where p_i is a piecewise linear ϵ_i-approximation to $p|C_i' \cup C_i''$, and neither γ_i' nor γ_i'' bounds mod 2 in $W_0 \cap$ int M_i, it follows that neither $p_i(\gamma_i')$ nor $p_i(\gamma_i'')$ bounds in V_0'. Thus $p_i(\gamma_i')$ is homologous mod 2 to $p_i(\gamma_i'')$ in V_0' and, for sufficiently large i, $g_i p_i(\gamma_i')$ is homologous mod 2 to $g_i p_i(\gamma_i'')$ in $W_0 \cap M_i$ and in $M_i - x_i$. Consequently, by Lemma 2.6* of [7], $\gamma_i' \sim \gamma_i''$ in $W_0 \cap M_i$ and $M_i - x_i$, for sufficiently large i. But, since x_i has coefficient 0 in this homology, then if $M_i - (C_i' \cup C_i'')$ is connected, every 3-cell in Γ_i has coefficient 0, which is impossible. Hence $M_i - (C_i' \cup C_i'')$ is not connected and for sufficiently large i all but one of the components of $M_i - (C_i' \cup C_i'')$ lie in W_0. Denote the union of C_i', C_i'' and these components by V_i. It may be assumed, as a consequence of the 0-regularity of the convergence of $\{M_i\}$ to M_0, that the sequence $\{V_i\}$ converges to V_0 and that $\{M_i - \text{int } V_i\}$ converges to $M_0 -$ int V_0.

Suppose that V_i has only one boundary surface. This must be C_i', since otherwise, the 0-regularity of the convergence of $\{M_i\}$ to M_0 implies that $\{V_i\}$ converges to C_0. However, it has been seen that, since γ_i' does not bound mod 2 in W_0, C_i' does not bound a 3-manifold with boundary in W_0. Hence bdry V_i has a component other than C_i'. If this is C_i'', then, since $M_i -$ int V_i is connected, there is an arc t_i from C_i'' to C_i' in $M_i -$ int V_i. Then $p_i(t_i)$ contains an arc t_0 from C_0 to C_0' which, for sufficiently large i, does not intersect the surface in V_0 corresponding to $C_0 \times (1/2)$ under the homeomorphism of V_0 onto $C_0 \times I$. This is impossible. Thus, the other boundary surfaces of V_i are components of K_i. Since this argument can be applied to each component of K_0, there is only one such boundary surface, C_i. Denote the non-trivial 2-cycle mod 2 carried by C_i by γ_i.

The cycle γ_i does not bound in $W_0 \cap M_i$. Consequently γ_i does not bound in $U_0 \cap M_i$. Also, $\{C_i\}$ converges to C_0 for otherwise $q_i(\gamma_i)$, where q_i is a piecewise linear ϵ_i-approximation to $q|C_i$, bounds in C_0 for sufficiently large i, $g_i q_i(\gamma_i)$ bounds in $U_0 \cap M_i$ and hence, applying Lemma 2.6* of [7], γ_i bounds in $U_0 \cap M_i$ for sufficiently large i. This contradiction completes the proof of the theorem.

In the Lemmas 3.2-3.5 and Theorem 3.6, C_0 denotes a component of K_0 and $\{C_i\}$ is a sequence converging strongly to C_0 such that for each i, C_i is a component of K_i.

<u>Lemma 3.2.</u> If J_0 is a simple closed curve in C_0, then there is a sequence $\{J_i\}$ of simple closed curves converging strongly to J_0 such that for each i, J_i lies in C_i.

<u>Proof.</u> This is a direct consequence of an application of Lemma 2.5* of [7] to the projection mapping of a regular neighborhood in E^n of C_0 onto C_0.

<u>Lemma 3.3.</u> If $\{J_i\}$ is a sequence of simple closed curves, as above, converging strongly to the simple closed curve J_0 in C_0 and for each i, γ_i is a non-trivial generating 1-cycle carried by J_i such that γ_i bounds in C_i (mod 2 or over the integers), then some multiple of γ_0 bounds in C_0.

<u>Proof.</u> Let U_0 denote a regular neighborhood of J_0 in E^n with projection mapping p and let A_0 denote a regular neighborhood of J_0 in $C_0 \cap U_0$. Then for sufficiently large i, $p_i(J_i) \subset A_0$, where p_i is a piecewise linear ϵ_i-approximation to $p|J_i$, and $p_i(\gamma_i)$ bounds in C_0. But $p_i(\gamma_i)$ is a multiple of γ_0.

<u>Corollary.</u> If $J_{01}, J_{02}, \ldots, J_{0m}$ are simple closed curves in C_0 carrying fundamental generating 1-cycles $\gamma_{01}, \ldots, \gamma_{0m}$ and if for each j, j = 1,..., m, there is a sequence $\{J_{ij}\}$ of simple closed curves converging strongly to J_{0j} such that for each i, J_{ij} lies in C_i and carries a fundamental 1-cycle γ_{ij} and there exist non-zero integers $b_{i1}, \ldots, b_{im}$ such that $\sum_j b_{ij} \gamma_{ij}$ bounds in C_i (mod the integers or mod 2) then there are non-zero integers $b_{01}, \ldots, b_{0m}$ such that $\sum b_{0j} \gamma_{0j}$ bounds in C_0.

If C_0 is orientable, $H_1(C_0)$ is generated by k independent cycles, $\gamma_{01}, ..., \gamma_{0k}$, of infinite order carried by k simple closed curves $J_{01}, ..., J_{0k}$. The integer k is even and C_0 is a 2-sphere with $k/2$ handles. If C_0 is non-orientable, $H_1(C_0)$ is generated by k independent 1-cycles of infinite order, $\gamma_{01}, ..., \gamma_{0k}$, and one, γ_{0k+1}, of order 2. Then C_0 is a sphere with $k + 1$ cross caps and k need not be even. It is seen from the above corollary that for sufficiently large i, $H_1(C_i)$ has at least as many independent genera-tors of infinite order as does $H_1(C_0)$ and that, since K_i is homeomorphic to K_0 and the above statements hold for <u>each</u> component of K_0, $H_1(C_0)$ has the same maximum number of independent generators as does $H_1(C_i)$ and the latter may be taken as the γ_{ij}, $j = 1, ..., k$, as described in the statement of the corollary. If C_i is non-orientable, denote by γ_{ik+1} a 1-cycle of order 2, which, together with $\{\gamma_{ij}\}_{j=1}^{k}$, generates $H_1(C_i)$.

Lemma 3.4. If B_0 is a disc with boundary J_0 such that $B_0 - J_0 \subset$ int M_0 and $J_0 \subset K_0$, then there is a sequence $\{B_i\}$ of discs with boundaries $\{J_i\}$ converging strongly to B_0 such that for each i, $B_i - J_i \subset$ int M_i and $J_i \subset K_i$.

Proof. This is essentially Lemma 2.13 of [7].

Lemma 3.5. If, in the statement of Lemma 3.2, γ_i denotes a fundamental generating 1-cycle carried by J_i, then if γ_0 bounds (over the integers) in C_0, then for sufficiently large i either γ_i or $2\gamma_i$ bounds over the integers in C_i.

Proof. If neither of these situations holds, then $\gamma_i \sim \sum_j a_{ij}\gamma_{ij}$, where the symbol "$\sim$" means "is homologous to", and at least one a_{ij} with $j < k + 1$ is not 0. Thus either $\gamma_i - \sum_{j=1}^{k} a_{ij}\gamma_{ij}$ or $2\gamma_i - \sum_{j=1}^{k} 2a_{ij}\gamma_{ij}$ is homologous to 0 in C_i. However, it follows from the corollary to Lemma 3.3 that there are integers c and $b_{01}, ..., b_{0k}$, c and at least one b_{0j} being different from 0, such that $c\gamma_0 - \sum_{j=1}^{k} b_{0j}\gamma_{0j} \sim 0$. Since the γ_{0j} are independent, this is impossible.

Theorem 3.6. The sequence $\{C_i\}$ converges completely regularly to C_0.

Proof. Suppose that P_0 is a point of C_0 in the interior of a disc D_0 in C_0 with boundary J_0 and that the J_{0j} have been adjusted so that they do not meet D_0. Let $B_0 = cl\,(C_0 - D_0)$. There is a sequence $\{J_i\}$ of simple closed curves converging strongly to J_0 such that for each i, J_i lies in C_i. There is a disc E_0 in M_0 such that $J_0 = E_0 \cap K_0$ and $E_0 \cup D_0$ bounds a 3-cell F_0 in M_0 and there is a sequence $\{E_i\}$ of discs converging strongly to E_0 such that for each i, $E_i \cap K_i = J_i$ and $E_i \subset M_i$. For each i, denote by γ_i a fundamental generating 1-cycle carried by J_i.

If γ_i bounds over the integers in C_i for sufficiently large i, J_i separates C_i into two sets whose closures may be denoted by B_i and D_i. Denote by x_i and y_i the non-trivial 2-cycles mod 2 carried by $D_i \cup E_i$ and $B_i \cup E_i$. Then x_i and y_i generate $H_2(D_i \cup E_i \cup B_i)$ mod 2. If W_0 is a regular neighborhood of $C_0 \cup E_0$ in E^n and P_i is a piecewise linear ϵ_i-approximation to the projection mapping of W_0, restricted to $W_0 \cap M_i$, then for sufficiently large i, $p_i(y_i) \sim a_i x_0 + d_i y_0$ mod 2 in $W_0 \cap M_0$ and, since x_0 bounds mod 2 in F_0, $p_i(y_i) \sim d_i y_0$ mod 2 in $(W_0 \cap M_0) \cup F_0$. Similarly, $p_i(x_i) \sim b_i y_0$ in $(W_0 \cap M_0) \cup F_0$. If one of b_i and d_i -- say b_i -- is 0, then $p_i(x_i)$ bounds mod 2 in M_0 and thus, since $x_i \sim g_i p_i(x_i)$ mod 2 in $W_0 \cap M_i$ for sufficiently large i, x_i bounds mod 2 in M_i. If both b_i and d_i are 1, then $p_i(x_i) \sim p_i(y_i)$ in $(W_0 \cap M_0) \cup F_0$ and thus $x_i (\sim g_i p_i(x_i))$ is homologous to y_i $(\sim g_i p_i(y_i))$ mod 2 in a proper subset of M_i. Thus $x_i + y_i \sim 0$ mod 2 on a proper subset of M_i. But $x_i + y_i$ is a non-trivial 2-cycle carried by C_i and cannot bound in a proper subset of M_i.

This contradiction shows that one of x_i and y_i--say x_i-- bounds mod 2 in M_i. Since, in this homology, each 3-simplex having a 2-face in B_i carries the coefficient 0, whereas each 3-simplex having a 2-face in D_i carries the coefficient 1, E_i separates M_i. The 0-regularity of the convergence of $\{M_i\}$ to M_0 implies that one of the closures of the components of $M_i - E_i$ may be labelled F_i in such a way that $\{F_i\}$ converges to F_0 and $\{M_i - F_i\}$ converges to $cl(M_0 - F_0)$ and that B_i and D_i may be so chosen that $\{B_i\}$ converges to B_0 and $\{D_i\}$ to D_0.

If ϵ is a positive number, D_0 and E_0 may be so chosen that $D_0 \cup E_0 \cup F_0$ has diameter less than ϵ. Thus, if $\{P_i\}$ and $\{Q_i\}$ are sequences of points converging to P_0 such that for each i, $P_i \cup Q_i \subset C_i$, $P_i \cup Q_i$ lies on an arc in D_i and D_i has diameter less than ϵ. Thus, if γ_i bounds in C_i for sufficiently large i, then the convergence of $\{C_i\}$ to C_0 is 0-regular at P_0.

If for sufficiently large i, γ_i does not bound in C_i, then $2\gamma_i$ bounds in C_i (Lemma 3.5). Denote by J_0' a simple closed curve in int D_0, by $\{J_i'\}$ a sequence of simple closed curves converging strongly to J_0' such that for each i, $J_i' \subset C_i$, and by $\{\gamma_i'\}$ a sequence of 1-cycles such that for each i, γ_i' generates $H_1(J_i')$. If γ_i' bounds in C_i for sufficiently large i or if either J_i or J_i' separates C_i, then the above argument demonstrates the 0-regularity of the convergence of $\{C_i\}$ to C_0 at P_0.

Suppose that γ_i' does not bound in C_i but that $2\gamma_i'$ does and that neither J_i nor J_i' separates C_i. Then $\gamma_i \sim \gamma_i'$ in C_i. It follows that $J_i \cup J_i'$ separates M_i and that $M_i - (J_i \cup J_i')$ has only two components. Denote the closures of these components by B_i and A_i. Denote by A_0 the annulus in D_0 bounded by $J_0 \cup J_0'$, by D_0' the closure of $D_0 - A_0$, by E_0' a disc with boundary J_0' such that $E_0' - J_0' \subset \text{int } F_0$, by F_0' the 3-cell in F_0 bounded by $D_0' \cup E_0'$ and by U_0 the closure of $F_0 - F_0'$. There is a sequence $\{E_i'\}$ of discs converging to E_0' such that for each i, bdry $E_i' = J_i'$ and $E_i' - J_i'$ lies in int $M_i - E_i$. Denote by x_0 and y_0' the non-trivial 2-cycles mod 2 carried by $E_0 \cup A_0 \cup E_0'$ and $D_0' \cup E_0'$ and by x_i and y_i those carried by $E_i \cup A_i \cup E_i'$ and $E_i \cup B_i \cup E_i'$. If W_0 is a regular neighborhood of $C_0 \cup E_0 \cup E_0'$ and p_i is a piecewise linear ϵ_i-approximation to the restriction to $W_0 \cap M_i$ of the projection mapping of W_0, then for sufficiently large i, $p_i(y_i) \sim a_i x_0 + d_i y_0 + d_i' y_0'$ mod 2 in $W_0 \cap M_0$ and, since x_0 and y_0' bound in F_0, $p_i(y_i) \sim d_i y_0$ in $F_0 \cup (W_0 \cap M_0)$. Similarly, $p_i(x_i) \sim b_i y_0$ in $F_0 \cup (W_0 \cap M_0)$. A repetition of the argument for the case where γ_i bounds in C_i now demonstrates that $M_i - (E_i \cup E_i')$ has two components and that the closure of one of them may be labelled U_i so that $\{U_i\}$ converges to U_0 and $\{M_i - U_i\}$ converges to $\text{cl}(M_0 - U_0)$. Since $M_i - U_i$ is connected and $M_0 - U_0$ is not, a contradiction arises. Hence for sufficiently large i, either γ_i or γ_i' bounds in C_i or J_i or J_i' separates C_i and it follows as above that the convergence of $\{C_i\}$ to C_0 is 0-regular at P_0.

To see that C_i is homeomorphic to C_0 for sufficiently large i, suppose that C_0 is not orientable and that V_0 is a regular neighborhood of C_0 in E^n. Suppose further that C_i is orientable for sufficiently large i and denote by γ_i a non-trivial 2-cycle mod the integers carried by C_i. For sufficiently large i, $p_i(\gamma_i)$, where p_i is a piecewise linear ϵ_i-approximation to the restriction to $V_0 \cap M_i$ of the projection map of V_0 onto C_0, is homologous to 0 in C_0. Thus $g_i p_i(\gamma_i)$ bounds over the integers in $V_0 \cap M_i$. However, since $\gamma_i \sim g_i p_i(\gamma_i)$ in $V_0 \cap M_i$ and γ_i does not bound in a proper subset of a component of M_i, a contradiction arises. Hence C_i is non-orientable for sufficiently large i and it follows from the remarks proceeding the statement of Lemma 3.4 that C_i is homeomorphic to C_0 for sufficiently large i. Since K_i is homeomorphic to K_0, it follows that if C_0 is orientable, then C_i is orientable and homeomorphic to C_0 for sufficiently large i. Hence Lemma 3 of [6] may be applied to prove that the convergence of $\{C_i\}$ to C_0 is h-1-regular and completely regular.

4. **Regular mappings whose inverses are 3-manifolds.** In this section, denote by $\{\epsilon_i\}$ a sequence of positive numbers converging to 0, by $\{g_i\}$ a sequence of mappings such that for each i, g_i is a piecewise linear ϵ_i-mapping of M_0 onto M_i carrying int M_0 onto int M_i and K_0 homeomorphically onto K_i and by $\{g_i'\}$ a sequence of mappings such that for each i, g_i' is a piecewise linear ϵ_i-mapping of M_i onto M_0 carrying int M_i onto int M_0 and $g_i'|K_i = g_i^{-1}|K_i$.

Lemma 4.1. If S_0 is a 2-sphere in int M_0 bounding a 3-cell C_0 in int M_0, then there is a sequence $\{S_i\}$ of 2-spheres converging strongly to S_0 such that for each i, S_i lies in and separates M_i and the closure, C_i, of one of the components of $M_i - S_i$ is in int M_i, all its homology groups vanish and it is such that $\{C_i\}$ converges to C_0 and $\{M_i - C_i\}$ converges to $cl(M_0 - C_0)$.

Proof. Denote by C_0' and C_0'' two 3-cells in int M_0 such that $C_0' \subset$ int C_0 and $C_0 \subset$ int C_0''. Denote by U_0 the set $cl(C_0'' - C_0')$, by W_0 a regular neighborhood of $cl(U_0)$ in E^n, by p the natural projection mapping of W_0 on U_0 and by p_i a piecewise linear ϵ_i-approximation to $p|W_0 \cap M_i$. The proof of Theorem 3.1 may be repeated to prove the existence of sequences $\{S_i'\}$ and $\{S_i''\}$ of compact 2-manifolds converging strongly to bdry C_0' and bdry C_0'', each $S_i' \cup S_i''$ lying in int M_i. Denote by x_i' and x_i'' non-trivial 2-cycles mod 2 carried by S_i' and S_i''. The proof of Theorem 3.1 demonstrates that, although x_i' and x_i'' do not bound mod 2 in W_0, they do bound mod 2 on a proper subset of M_i. Thus S_i' and S_i'' bound 3-manifolds with boundary, C_i' and C_i'', in M_i. The 0-regularity of the convergence of $\{M_i\}$ to M_0 demonstrates that $\{C_i'\}$ and $\{C_i''\}$ converge to C_0' and C_0'', that $C_i' \subset$ int C_i'' for sufficiently large i, that $cl(C_i'' - C_i') = U_i$ is a compact 3-manifold with boundary and that $\{U_i\}$ converges to U_0 and $\{M_i - U_i\}$ converges to $cl(M_0 - U_0)$.

If γ_0 denotes a fundamental 2-cycle mod the integers carried by S_0, then for sufficiently large i, $p_i g_i(\gamma_0)$ bounds mod the integers in int C_0'' and thus $g_i p_i g_i(\gamma_0)$ bounds in $(W_0 \cap M_i) \cup C_i''$ and so does $g_i(\gamma_0)$. Thus $g_i(\gamma_0)$ bounds in int C_i'' and is, therefore, homologous mod the integers in U_i to a cycle on S_i' which is homologous to

0 in $C_i{}'$. However, such a cycle on $S_i{}'$ is not homologous to 0 on $S_i{}'$ for, if it were, $g_i(\gamma_0)$ would bound in U_i and $p_i g_i(\gamma_0)$ would bound in $W_0 \cap M_0$, which, since $p_i g_i | S_0$ is homotopic in W_0 to the identity map, is impossible. Thus $S_i{}'$ carries a non-bounding 2-cycle mod the integers and is orientable. A similar argument demonstrates that $S_i{}''$ carries a non-bounding 2-cycle γ'' and is orientable. Also, for sufficiently large i, $p_i(\gamma_i{}'')$ bounds in $(W_0 \cap M_0) \cup C{}''$ and thus $g_i p_i (\gamma_i{}'')$, which is homologous to $\gamma_i{}''$, bounds in $(W_0 \cap M_i) \cup C_i{}''$. Thus $\gamma_i{}''$ bounds in $C_i{}''$ and $C_i{}''$ is orientable. Consequently U_i and $C_i{}'$ are orientable.

Since S_0 is not contractible in W_0 and $p_i g_i | S_0$ is homotopic in U_0 to the identity on S_0, $g_i(S_0)$ is not contractible in U_i. Consequently it follows from the Sphere Theorem (Papakyriakopoulos [13] and Whitehead [20]) that there is a non-singular 2-sphere in U_i which is not contractible in U_i. Since U_0 was chosen arbitrarily, the existence of the required sequence $\{S_i\}$ is demonstrated. (If γ_i, a non-trivial 2-cycle carried by S_i, bounds in U_i, $p_i(\gamma_i)$ bounds in $U_0 \cup (W_0 \cap M_0)$ so that $p_i | S_i$ is homotopic to 0 in U_0. Then $g_i p_i | S_i$, which is, as a consequence of Lemma 2.6* of [7], homotopic to the identity in U_i, is homotopic to 0 in U_i -- a contradiction.) Also, the sphere S_i carries a non-trivial 2-cycle which bounds in $C_i{}''$. The arguments above and in the proof of Theorem 3.1 thus prove that $M_i - S_i$ has two components, the closure, C_i, of one of them being homologically trivial and orientable. The 0-regularity of the convergence of $\{M_i\}$ to M_0 implies that $\{C_i\}$ converges to C_0 and $\{M_i - C_i\}$ to $\mathrm{cl}(M_0 - C_0)$.

<u>Definition 4.2.</u> A simple closed curve in M_i which bounds a disc in M_i will be called <u>unknotted</u> in M_i. It also lies on a 2-sphere in M_i.

<u>Lemma 4.3.</u> If J_0 is an unknotted simple closed curve in int M_0, then there is a sequence $\{J_i\}$ of simple closed curves converging strongly to J_0 such that for each i, J_i is unknotted in M_i.

<u>Proof.</u> There is a 2-sphere S_0 bounding a 3-cell in int M_0 such that S_0 contains J_0 and there is a sequence $\{S_i\}$ of 2-spheres converging strongly to S_0 such that for each i, S_i lies in M_i. It follows from Lemma 2.5 of [7] (see also Lemma 3.2 above) that there is a sequence $\{J_i\}$ of simple closed curves converging strongly to J_0 such that for each i, J_i lies in S_i. This proves the lemma.

Lemma 4.4. If J_0 is a simple closed curve in int M_0 bounding a disc D_0 in int M_0 and $\{J_i\}$ is a sequence of simple closed curves converging strongly to J_0 such that for each i, J_i is unknotted in int M_i, then there is a sequence $\{D_i\}$ of discs converging to D_0 such that for each i, D_i lies in int M_i and is bounded by J_i.

Proof. Let S_0 denote a 2-sphere in int M_i which bounds a 3-cell C_0 whose interior contains J_0 and let $\{S_i\}$ be a sequence of 2-spheres converging strongly to S_0 such that for each i, $S_i \subset$ int M_i and bounds a homologically trivial 3-manifold with boundary C_i. Then $\{C_i\}$ converges to C_0 and $\{M_i - C_i\}$ converges to $cl(M_0 - C_0)$. Let D_i' denote a disc in M_i bounded by J_i, which, after small adjustments, may be assumed to lie in int M_i and to be such that each component of $D_i' \cap S_i$ is a simple closed curve. Clearly $J_i \subset$ int C_i for sufficiently large i.

Suppose D is a disc in S_i whose boundary, J, lies in D_i' but such that $D_i' \cap$ int D $= 0$. Then the disc in D_i' bounded by J may be replaced by D and moved slightly away from S_i so that a disc D_i'' is obtained which is bounded by J_i and is such that $D_i'' \cap S_i$ has fewer components than $D_i' \cap S_i$. This process may be repeated until a disc is obtained which lies in C_i and is bounded by J_i. Since C_0 was chosen arbitrarily, the existence of the required sequence is demonstrated.

In the sequel, S_0 will denote a 2-sphere in int M_0 bounding a 3-cell C_0, C_0' and C_0'' will denote 3-cells in int M_0 bounded by S_0' and S_0'' such that $C_0' \subset$ int C_0 and $C_0 \subset$ int C_0'' and U_0 will denote $cl(C_0'' - C_0')$. It follows from Lemma 4.1 that there are sequences $\{S_i'\}$, $\{S_i\}$, $\{S_i''\}$, $\{C_i'\}$, $\{C_i\}$, $\{C_i''\}$, and $\{U_i\}$ converging strongly to S_0', S_0, S_0'', C_0', C_0, C_0'', and U_0 such that $\{M_i - U_i\}$ converges to $cl(M_0 - U_0)$ and for each i, S_i', S_i, and S_i'' are 2-spheres bounding C_i', C_i, and C_i'' which are compact, orientable 3-manifolds with boundary in int M_i, $C_i' \subset$ int C_i, $C_i \subset$ int C_i'' and $cl(C_i'' - C_i') = U_i$. Furthermore, C_i', C_i, and C_i'' are homologically trivial.

Lemma 4.5. If W_0 is a regular neighborhood of S_0 in E_n, which, without loss of generality, may be assumed to contain U_0, and p is its projection map, then for sufficiently large i, $p_i|S_i$ is homotopic in S_0 to a homeomorphism of S_i into S_0, where p_i is a piecewise linear ϵ_i-approximation to $p|W_0 \cap M_i$.

Proof. Denote by γ_i a fundamental 2-cycle carried by S_i and by γ_0'' one carried by S_0''. It has been seen in the proof of Lemma 4.1 that $g_i(\gamma_0'')$, being homologous to 0 in $(W_0 \cap M_i) \cup C_i'$, is homologous in $W_0 \cap M_i$ to a non-zero multiple of γ_i-- say $k_i \gamma_i$ -- for sufficiently large i. But for sufficiently large i, $\gamma_0 \sim \gamma_0'' \sim p_i g_i(\gamma_0'')$, $p_i g_i(\gamma_0'') \sim k_i p_i(\gamma_i)$ and $p_i(\gamma_i) \sim n_i \gamma_0$ in $W_0 \cap M_0$, where n_i is a non-zero integer. Hence $\gamma_0 \sim k_i n_i \gamma_0$ and $k_i = n_i = \pm 1$. Hence $p_i | S_i$ has degree ± 1 and $p_i | S_i$ is homotopic to a homeomorphism.

$\underline{\text{Lemma 4.6}}$. Suppose that J_0 is a simple closed curve in S_0, V_0 is a regular neighborhood in E^n of J_0 consistently imbedded in a regular neighborhood W_0 in E^n of S_0, the closures of the components of $S_0 - J_0$ are denoted by D_0 and E_0 and for each i, h_i is a mapping of S_0 into U_i chosen in such a way that the sequences $\{h_i(D_0)\}$, $\{h_i(E_0)\}$ and $\{h_i(J_0)\}$ converge to D_0, E_0, and J_0. Then (1) $h_i | J_0$ is homotopic to 0 in V_0 for sufficiently large i if and only if $h_i | S_0$ is homotopic to 0 in U_i for sufficiently large i and (2) if $p_i h_i$, where p_i is a piecewise linear e_i-approximation to the restriction to $W_0 \cap M_i$ of the projection map of W_0 onto S_0, is homotopic to a homeomorphism of S_0 into itself, then $p_i h_i | J_0$ is homotopic to a homeomorphism in $V_0 \cap S_0$.

$\underline{\text{Proof}}$. Let U_D and U_E denote regular neighborhoods of D_0 and E_0 in E^n in each of which V_0 is consistently imbedded and which are consistently imbedded in W_0. Since $p_i h_i(D_0) \subset (V_0 \cap S_0) \cup D_0$ and $p_i h_i(E_0) \subset (V_0 \cap S_0) \cup E_0$, it follows from standard theorems that if F_i is a mapping of $J_0 \times I$ into $V_0 \cap S_0$ such that $F_i(x,0) = p_i h_i(x)$ for each x in J_0, then there is a mapping F_i^* of $S_0 \times I$ into S_0 which extends F_i and is such that $F_i^*(x,0) = p_i h_i(x)$ for each x in S_0 and $F_i^*(D_0,t) \subset D_0 \cup (V_0 \cap S_0)$ and $F_i^*(E_0,t) \subset E_0 \cup (V_0 \cap S_0)$. In particular, if $F_i(J_0 \times (1)) \subset J_0$, then F_i^* may be constructed so that the degrees of $F_i | J_0 \times (1)$ and $F_i^* | S_0 \times (1)$ are the same, orientation of generating cycles of $H_1(J_0)$ and $H_1(S_0)$ being appropriately chosen, Also, it follows from Lemma 2.6* of [7] that for sufficiently large i there is a mapping G_i of $J_0 \times I$ into $V_0 \cap M_i$ and an extension of G_i to a map G_i^* of $S_0 \times I$ into V_i such that $G_i^*(x,0) = h_i(x)$, $G_i^*(x,1) = g_i p_i h_i(x)$, and $G_i^*(D_0,t) \subset U_D$ and $G_i^*(E_0,t) \subset U_E$ for each t. A combination of the homotopies G_i^* and $g_i F_i^*$ yields a mapping H_i of $J_0 \times I$ into $V_0 \cap M_i$ and an extension of H_i to a map H_i^* of $S_0 \times I$ into U_i such that $H_i^*(x,0) = h_i(x)$, $H_i^*(x,1) = g_i F_i^*(x,1)$, $H_i^*(D_0,t) \subset U_D$ and $H_i^*(E_0,t) \subset U_E$ for each t.

The mapping $h_i | J_0$ is, for sufficiently large i, homotopic to 0 in V_0 if and only if

$p_i h_i | J_0$ is homotopic to 0 in $V_0 \cap S_0$ and the above remarks demonstrate that $p_i h_i | J_0$ is homotopic to 0 in $V_0 \cap S_0$ if and only if $p_i h_i$ is homotopic to 0 in S_0 which is true if and only if $g_i p_i h_i$ is homotopic to 0 in $W_0 \cap M_i$ which is true if and only if h_i is homotopic to 0 in U_i. The above remarks also demonstrate that if $p_i h_i$ has degree ± 1, so also does $p_i h_i | J_i$.

Lemma 4.7. If J_0 is a simple closed curve in S_0, then the sequence $\{S_i\}$ may be selected in such a way that there is a sequence $\{J_i\}$ of simple closed curves converging properly to J_0 such that for each i, $J_i \subset S_i$.

Proof. Let D_0 and E_0 denote the closures of the components of $S_0 - J_0$, let U_D and U_E denote regular neighborhoods of D_0 and E_0 in E^n whose common part is a regular neighborhood, V_0, of J_0 consistently imbedded in U_D and U_E and whose union is a regular neighborhood W_0 of S_0 in which U_D and U_E are consistently imbedded. There are sequences $\{D_i'\}$, $\{E_i'\}$ and $\{J_i'\}$ converging strongly to D_0, E_0, and J_0 such that for each i, D_i' and E_i' are discs in U_i with boundary J_i'. For sufficiently large i, $D_i' \subset U_D$, $E_i' \subset U_E$ and $D_i' \cap E_i' \subset V_0$ and small adjustments may be made so that each component of $D_i' \cap E_i'$ is a simple closed curve.

Since for sufficiently large i, J_i' is, by construction, not contractible in V_0, there is a disc D_i'' in D_i' whose boundary is a simple closed curve J_i which lies in E_i' and is not contractible in V_0 but which contains no other component of $E_i' \cap D_i'$ which is not contractible in V_0. Let E_i be the disc in E_i' bounded by J_i. If F is a disc in E_i which lies except for its boundary, which is a component of $(D_i'' \cap E_i) - J_i$, in $E_i - (D_i'' \cap E_i)$, then the disc in D_i'' bounded by bdry F may be replaced by F and moved slightly away from E_i to form a disc with boundary J_i whose common part with E_i has fewer components than $D_i'' \cap E_i$. A repetition of this process yields a disc D_i such that $D_i \cap E_i = J_i = $ bdry D_i. Let S_i denote the 2-sphere $D_i \cup E_i$. Clearly $\{J_i\}$ converges strongly to J_0 and $\{E_i\}$ converges to E_0.

Let h_i be a homeomorphism of S_0 into S_i carrying J_0 onto J_i, D_0 onto D_i and E_0 onto E_i. Then h_i is homotopic in U_i, for sufficiently large i, to a mapping h_i' taking D_0 into U_D, $h_i(E_0)$ remaining pointwise fixed under this homotopy. To see this, note that each component of $D_i - (D_i \cap U_D)$ lies in a disc F_i in $D_i \cap U_E$ whose boundary, E_i, is contractible in V_0 and lies in a $2\epsilon_i$-neighborhood of J_0. Let F_i'' be a disc in D_i whose interior contains F_i and whose boundary is contractible in V_0 and lies together with $F_i'' - F_i$ in a $2\epsilon_i$-neighborhood of J_0 and let F_i' be a disc in int F_i'' with boundary B_i' such that $F_i \subset$ int F_i'. There is a mapping T_i of $h_i^{-1}(B_i) \times I$ into $V_0 \cap S_0$ such that $T_i(x,0) = p_i h_i(x)$ for x in $h_i^{-1}(B_i)$ and $T_i(x,1) = P_0$, a point of V_0. Then T_i may be extended to a mapping T_i^* of $h_i^{-1}(F_i) \times I$ into $(V_0 \cap S_0) \cup E_0$ such that $T_i^*(x,0) = p_i h_i(x)$ and $T_i^*(x,1) = P_0$ for each x in $h_i^{-1}(F_i)$ and this may be extended to a mapping T_i^{**} of $h_i^{-1}(F_i') \times I$ into S_0 such that $T_i^{**}(x,0) = p_i h_i(x)$ for x in $h_i^{-1}(F_i')$, $T_i^{**}(x,t) = p_i h_i(x)$ for x in $h_i^{-1}(B_i')$ and $T_i^{**}(x,t)$ is a point of V_0 for x in $F_i' - F_i$ and all t. The map $h_i|h_i^{-1}(F_i')$ is homotopic to $g_i p_i h_i|h_i^{-1}(F_i')$ in U_E. Thus there is a mapping G_i of $S_0 \times I$ into U_i such that $G_i(x,t) = h_i(x)$ for x in $S_0 - h_i^{-1}(F_i'')$, $G_i(x,0) = h_i(x)$ for x in S_0, $G_i(x,1) = g_i p_i h_i(x)$ for x in $h_i^{-1}(F_i')$. But $g_i T_i^{**}$ may be extended to a mapping H_i of $S_0 \times I$ into U_i such that $H_i(x,t) = h_i(x)$ for x in $S_0 - h_i^{-1}(F_i'')$ and $H_i(x,0) = G_i(x,1)$. A combination of G_i and H_i yields a homotopy of h_i under which $h_i(x)$ remains unchanged for x outside a small neighborhood of $h_i^{-1}(F_i)$ and which carries F_i into $V_0 \cap M_i$. A repetition of this process yields the required homotopy of h_i to h_i'.

Since h_i is a homeomorphism, it follows from Lemma 4.5 that $p_i h_i'$ is homotopic to a homeomorphism and from Lemma 4.6 that $p_i h_i'|J_0 = p_i h_i|J_0$ is homotopic to a homeomorphism in V_0. Thus $p_i|J_i$ is homotopic to a homeomorphism in V_0 and the lemma is proved.

Since C_i'' is orientable and homologically trivial, it makes sense to consider intersection numbers and linking numbers of pairs of 1-cycles in C_i''. See [2] and [16].

Lemma 4.8. If z_0 is an unknotted simple closed curve in int C_0'', ϵ is a positive number and $\{z_i\}$ is a sequence of simple closed curves converging strongly to z_0 such that for each i, z_i is unknotted in C_i'', then there is an integer N such that if $n > N$ and z_0' is a closed curve in int C_0'' which links z_0 and whose distance from z_0 exceeds ϵ, then $g_n(z_0')$ links z_n.

Proof. It may be supposed that S_0, as defined previously, is a 2-sphere in int C_0'' containing z_0, that D_0 and E_0 are the closures of the components of $S_0 - z_0$, and that V_0, W_0, U_D and U_E are as described in the proof of Lemma 4.7 with the additional properties that each is an $\epsilon/6$-neighborhood and that $W_0 \subset$ int W_0', where W_0' is a regular $\epsilon/6$-neighborhood of S_0 such that $M_0 \cap$ bdry W_0' consists of two 2-spheres Q_0' and Q_0'', Q_0' is in int C_0, and Q_0'' is in ext C_0. Denote by q the projection map of V_0 on z_0. Let N be such that for $n > N$, $\epsilon_n < d(Q_0' \cup Q_0'', S_0)/6$ and let z_0' be a closed curve in int C_0'' whose distance from z_0 exceeds ϵ and which links z_0. Then z_0' intersects both ext Q_0'' and int Q_0' (interiors being the components of the complement lying in C_0'').

It follows from Lemma 4.7 that there are a sequence $\{S_i\}$ of 2-spheres converging to S_0, a sequence $\{z_i''\}$ of simple closed curves converging properly to z_0 and sequences $\{E_i\}$ and $\{D_i\}$ of discs converging to E_0 and a subset of S_0 respectively such that for each i, $z_i'' \subset S_i$, E_i and D_i are the closures of the components of $S_i - z_i''$ and D_i is deformable in $W_0 \cap M_i$, for sufficiently large i, into a subset D_i' of U_D under a deformation leaving z_i'' pointwise fixed. The construction may be carried out so that $\{D_i\}$ converges to D_0. The integer N should also be chosen so that for $n > N$, $\epsilon_n < d[g_n(Q_0' \cup Q_0''), S_n \cup D_n']/6$, $q_n|z_n''$ is homotopic to a homeomorphism in z_0, $g_n q_n|z_n \cup z_n''$ is homotopic to the identity in $V_0 \cap M_n$ and $W_0 \cap g_n(Q_0' \cup Q_0'') = C$, where q_n is a piecewise linear ϵ_n-approximation to $q|V_0 \cap M_n$.

Fix $n > N$, let z_n' denote a simple closed curve which is the image of z_0' under a very close homeomorphic approximation g_n'' to $g_n|z_0'$ and let γ_n, γ_n', and γ_n'' denote fundamental 1-cycles carried by z_n, z_n', and z_n''. If z_n' links z_n, so does $g_n(z_0')$. Also, $q_n(\gamma_n'') \sim \gamma_0$ in z_0 and therefore $q_n(\gamma_n) \sim k_n q_n(\gamma_n'')$ in z_0, where k_n is a non-zero integer. Thus $\gamma_n \sim g_n q_n(\gamma_n) \sim k_n g_n q_n(\gamma_n'') \sim k_n \gamma_n''$ in $V_0 \cap M_n$. Denote by T_n the boundary of a solid torus V_n in $C_n'' - (z_n \cup z_n'')$ which contains z_n' in its interior and whose first homology group is generated by γ_n'. Since C_n'' is homologically trivial, γ_n'' is homologous in $C_n'' -$ int T_n to a cycle a_n on T_n which is homologous to 0 in V_n and $k_n \gamma_n''$ is homologous to $k_n a_n$. Thus, if γ_n'' links γ_n', a_n links γ_n' and is not homologous to 0 in T_n. Then $k_n a_n$ is not homologous to 0 in T_n and, again since C_n'' is homologically trivial, $k_n a_n$ links γ_n'. Therefore, γ_n'' and γ_n each link γ_n' and the lemma will be proved if it is demonstrated that γ_n'' links γ_n'.

Let Γ denote the collection of components of $(Q_0'' \cup$ int $Q_0'' -$ int $Q_0') \cap z_0'$ which intersect Q_0'', D_0 and Q_0' and let Γ^* denote the union of the elements of Γ. The collection Γ exists because z_0' links z_0 and has distance from z_0 exceeding ϵ. Denote by

K'' the set of endpoints of elements of Γ which lie in Q_0'' and by K' those which lie in Q_0'. Consider $g_n(\Gamma^*)$. The set $g_n(K'')$ lies in ext S_n and $g_n(K') \subset$ int S_n and they carry 0-cycles, a_n'' and a_n' such that $a_n'' - a_n'$ bounds a 1-chain x_n carried by $g_n(\Gamma^*)$. The linking number, $|v(z_n', z_n'')|$ equals the intersection number, $|\phi(x_n, D_n')|$. This number is $|\phi(x_n, D_n')|$. To see this, note that if $p \in z_0'$ and $g_n(p) \in D_n'$, then either $p \in \Gamma^*$ or p is a point of a component t of $(Q_0'' \cup$ int Q_0'' - int $Q_0') \cap z_0'$ whose endpoints a and b lie on the same one of Q_0' and Q_0'' -- say Q_0''. Then $g_n(a \cup b) \subset$ ext S_0 and $E_n \cap g_n(t) = 0$. Thus $|\phi(g_n(t), D_n')| = 0$. However, if z_n' does not link z_n'', then $0 = |\phi(z_n', D_n')| = |\phi(x_n, D_n')|$ $= |\phi(x_n, D_n' \cup E_n)| = |\phi(x_n, S_n)|$. This contradicts the fact that S_n separates the carriers of a_n' and a_n'' in C_n''. Thus z_n' and z_n'' and, consequently, $g_n(z_0)$ and z_n are linked.

In the sequel and for brevity a symbol with the single subscript, i, will ordinarily denote a point set in C_i'' as will a symbol with a double subscript whose first element is i. The exceptions to this rule will not cause confusion.

<u>Definition 4.9.</u> If the annulus A in M_i lies on a disc in M_i, then A is said to be <u>unknotted</u> in M_i.

<u>Theorem 4.10.</u> If A_0 is an unknotted annulus in int C_0'', then there is a sequence A_i of annuli converging strongly to A_0 such that for each i, A_i is unknotted in int C_i''.

<u>Proof.</u> Denote by J_0 a boundary curve of A_0 and by D_0 a disc in C_0'' which contains A_0 and is bounded by J_0. Let $U_{00}, U_{01}, \ldots, U_{05}$ denote solid tori in C_0'' such that $J_0 \subset$ int U_{00}, $U_{00} \subset$ int $U_{01}, \ldots, U_{04} \subset$ int U_{05}, $A_0 \subset$ int U_{05}, and $A_0 \cap$ bdry U_{0j} is a simple closed curve J_{0j} in int A_0 for $j = 0, \ldots, 4$. Let $Q^1, \ldots, Q^7$ denote regular neighborhoods in E^n of $U_{04} \cap A_0$ such that $U_{00} \subset$ int Q^1, $Q^1 \subset$ int $Q^2, \ldots, Q^6 \subset$ int Q^7, $Q^7 \cap M_0$ $\subset C_0'' \cap$ int U_{05}, Q^7 does not contain bdry $A_0 - J_0$, and $Q^k \cap$ bdry U_{0j} is an annulus for $k = 1, \ldots, 7$ and $j = 1, \ldots, 4$. For $j = 0, \ldots, 5$, let $V_j^1, \ldots, V_j^5$ denote regular neighborhoods in E^n of bdry U_{0j} such that $V_j^1 \subset$ int $V_j^2, \ldots, V_j^4 \subset$ int V_j^5, $V_j^5 \cap M_0$ $\subset C_0'' \cap ($int $U_{0j+1} - U_{0j-1})$, $V_j^5 \cap V_k^5 = 0$ unless $j = k$, $V_5^5 \cap (Q^7 \cup A_0) = 0$, $V_0^5 \subset$ int Q^1, $Q^1 \cap [U_{05} - U_{05} \cap (U_{04} \cup V_4^5)] \neq 0$, and $V_j^p \cap Q^k$ is a regular neighborhood of $Q^k \cap$ bdry U_{0j} for $p = 1, \ldots, 5$, $j = 0, \ldots, 4$ and $k = 1, \ldots, 7$. Let W_0 be a regular neighborhood in E^n of D_0 such that $Q^7 \cup U_{04} \cup V_4^5 \subset W_0$ and W_0 - int U_{05} projects into $D_0 - A_0$ under the natural projection mapping of W_0 onto D_0. It follows from the proofs of Theorem 3.1 and Lemma 4.1 that there is, for each j, a sequence $\{U_{ij}\}$ of

compact 3-manifolds with boundary converging strongly to U_{0j} where $U_{i0} \subset \text{int } U_{i1}$, $U_{i1} \subset \text{int } U_{i2}$ et cetera. Also, there are a sequence $\{J_i\}$ of simple closed curves converging strongly to J_0 and a sequence $\{D_i\}$ of discs such that for each i, $J_i = \text{bdry } D_i$ and for sufficiently large i, $D_i \subset \text{int } W_0 \cap C_i''$. It may be assumed that $J_i \subset \text{int } U_0$ and that each component of $D_i \cap \text{bdry } U_{ij}$ is a simple closed curve. Note that is is not asserted that $\{D_i\}$ converges to D_0.

An essential component of $D_i - (D_i \cap \text{int } U_{i5})$ is such a component D which is not deformable in a regular neighborhood of $\text{cl}(D_0 - A_0)$ into a regular neighborhood of U_{i5} under a deformation which moves no point of bdry D. Denote the number of such components by $k(D_i)$. It may be supposed that if D_i' is any disc in $\text{int } W_0 \cap C_i''$ which is bounded by J_i and is such that each component of $D_i' \cap U_{ij}$ is a simple closed curve, then $k(D_i) \leqq k(D_i')$. Since J_i is not contractible in U_{i0}, the proofs of Lemmas 4.6 and 4.7 imply that $k(D_i) > 0$.

Denote by N_i a component of $(U_{i4} - \text{int } U_{i0}) \cap D_i$. The set N_i is a disc with holes; i.e. the topological image of a circular disc from which have been removed the interiors of finitely many circles. For sufficiently large i, $g_i'(N_i \cap \text{bdry } U_{i4})$ and $g_i g_i'(N_i \cap \text{bdry } U_{i4})$ are subsets of V_4^1. Also, no component, x, of $N_i \cap \text{bdry } U_{i4}$ links J_i so, by Lemma 4.8, $g_i'(x)$ does not link J_0. Hence $g_i'(N_i \cap \text{bdry } U_{i4})$ is deformable in $V_4^1 \cap C_0''$ into a subset of Q^1. Therefore, for sufficiently large i, $g_i g_i'(N_i \cap \text{bdry } U_{i4})$ is deformable in $V_4^2 \cap C_i''$ into a subset of Q^2. Hence there is a mapping F_i of $(N_i \cap \text{bdry } U_{i4}) \times I$ into $V_4^3 \cap C_i''$ such that $F_i(x,0) = x$ and $F_i(x,1) \in Q^3$. The set $F_i[(N_i \cap \text{bdry } U_{i4}) \times I]$ is the union of singular annuli. For $p \in N_i \cap \text{bdry } U_{i4}$, denote $F_i(p,1)$ by $f_i(p)$. For sufficiently large i, $g_i'(N_i \cap \text{bdry } U_{i0})$ and $g_i g_i'(N_i \cap \text{bdry } U_{i0})$ are subsets of V_0^1 and $g_i'f_i(N_i \cap \text{bdry } U_{i4})$ and $g_i g_i'f_i(N_i \cap \text{bdry } U_{i4})$ are subsets of $Q^4 \cap V_4^4$. It follows, as usual, from the h-2-regularity and Lemma 2.6* of [7], that, since $N_i \cap \text{bdry } U_{i0}$ and $f_i(N_i \cap \text{bdry } U_{i4})$ bound a singular disc with holes in $C_i'' \cap ([(V_4^3 \cup U_{i4}) - U_{i0}] \cup V_0^1)$, then $g_i'(N_i \cap \text{bdry } U_{i0})$ and $g_i'f_i(N_i \cap \text{bdry } U_{i4})$ bound a singular disc with holes in $C_0'' \cap ([(V_4^4 \cup U_{04}) - U_{00}] \cup V_0^2)$ and consequently these sets bound a singular disc with holes in $C_0'' \cap (Q^4 - [Q^4 \cap (U_{00} - V_0^2)])$. Thus $g_i g_i'(N_i \cap \text{bdry } U_{i0})$ and $g_i g_i'f_i(N_i \cap \text{bdry } U_{i4})$ bound a singular disc with holes L_i in $C_i'' \cap (Q^5 - [Q^5 \cap (U_{i0} - V_0^3)])$. Also, $N_i \cap \text{bdry } U_{i0}$ and $g_i g_i'(N_i \cap \text{bdry } U_{i0})$ bound a union of singular annuli, L_{i0} in $V_0^2 \cap Q^4$ and $f_i(N_i \cap \text{bdry } U_{i4})$ and $g_i g_i' f_i(N_i \cap \text{bdry } U_{i4})$ bound a union of singular annuli, L_{i4} in $Q^5 \cap V_4^5$. (The exact details involved in this argument are quite like those in the proofs of Lemmas 4.6 and 4.7 and are, for that reason, not repeated here.)

A singular disc D_i' may now be constructed by piecing together $U_{i0} \cap D_i$, $D_i - (U_{i4} \cap D_i)$, $F_i[(N_i \cap \text{bdry } U_{i4}) \times I]$, L_{i0}, L_{i4} and L_i. Dehn's Lemma [13] may now be applied to obtain a disc E_i in $W_0 \cap C_i''$ whose boundary is J_i and which is such that $k(E_i) \leqq k(D_i)$ and $E_i \cap U_{i3} \subset Q^5$. (The proof of Dehn's Lemma constructs E_i in such a way that E_i lies on D_i' except in arbitrarily small neighborhoods of the double lines and triple points of D_i'.)

For sufficiently large i, at least one of the components of $E_i \cap \text{bdry } U_{i1}$ is a simple closed curve which is not contractible in V_1^1 for, otherwise, the proofs of Lemmas 4.6 and 4.7 demonstrate that E_i may be replaced by a disc in $U_{i1} \cup V_1^5$ whose boundary is J_i, which, by the construction of J_i, is impossible. Let B_i be a disc in E_i bounded by a simple closed curve x_i in $E_i \cap \text{bdry } U_{i1}$ such that x_i is not contractible in V_1^1 but every other component of $B_i \cap \text{bdry } U_{i1}$ is contractible in V_1^1. For sufficiently large i, each component of $B_i \cap \text{bdry } U_{i3}$ is contractible in a set Z_i in a regular neighborhood $W_0' \subset W_0$ of $\text{cl}[D_0 - (D_0 \cap U_{03})]$ which does not intersect $U_{i2} \cup V_2^5$. Thus $(B_i \cap U_{i3}) \cup Z_i$ is a singular disc which, by Dehn's Lemma, may be replaced by a non-singular disc B_i' which is bounded by x_i, which is such that $B_i' \cap (U_{i2} \cup V_2^5) \subset B_i \cap (U_{i2} \cup V_2^5)$ and which lies in $(Q^5 \cup W_0') \cap C_i''$. In C_0'', J_{01} $(= U_{01} \cap A_0)$ bounds a disc D_0' such that $D_0' \cap D_0 = J_{01}$ and the various regular neighborhoods described here could have been chosen so that $D_0' \cap W_0' = 0$ and $D_0' \cap Q^5 \subset V_1^1$. An application of Lemmas 4.4, 4.6 and 4.7 and their proofs demonstrates that there is, for sufficiently large i, a disc D_i' in $C_i'' - (C_i'' \cap W_0')$ which is bounded by x_i and is such that $D_i' \cap Q^6 \subset V_1^2$. Thus $D_i' \cap B_i' \subset Q^5 \cap V_1^2$. An application of the proofs of Lemmas 4.6 and 4.7 now demonstrates that for sufficiently large i, there is in $B_i' \cap V_1^2$ and, consequently, in $B_i \cap V_1^2$ and $E_i \cap V_1^2 \cap Q^5$ a simple closed curve J_{i1} such that $g_i'|J_{i1}$ is homotopic in $V_1^3 \cap Q^6$ to a homeomorphism of J_{i1} into J_{01}. A similar argument proves that there is a simple closed curve J_{i2} in B_i and $E_i \cap V_2^2 \cap Q^5$ such that $g_i'|J_{i2}$ is homotopic in $V_2^3 \cap Q^6$ to a homeomorphism of J_{i2} into J_{02}.

The simple closed curves J_{i1} and J_{i2} bound an annulus H_i in E_i. For sufficiently large i, J_{i1} and J_{i2} bound a singular annulus H_i' in $(Q^7 \cap C_i'') - [J_i \cup (W_0' \cap Q^7 \cap C_i'')]$. This follows from an application of the arguments in the proofs of Lemmas 4.6 and 4.7 and the facts that, by the construction of J_{i1} and J_{i2}, $g_i'(J_{i1})$ and $g_i'(J_{i2})$ bound a singular annulus in $(Q^6 \cap C_0'') - [(U_{00} - V_0^4) \cup (W_0' \cap Q^6 \cap C_0'')]$. If H_i contains an essential component of $E_i - (E_i \cap U_{i5})$, replace H_i in E_i by H_i' and apply Dehn's Lemma to the resulting singular disc in order to obtain a disc E_i' in $W_0 \cap C_i''$ bounded by J_i such that $k(E_i') < k(E_i)$. This contradiction implies that H_i contains no such essential components. Hence the arguments in the proofs of Lemmas 4.6 and 4.7 may again be applied to obtain a deformation of H_i into a singular annulus H_i' in U_{i4} under which no point of U_{i3} is moved and no point is moved into U_{i3}. Thus the extension of Dehn's Lemma to discs with holes [18] implies the existence in U_{i4} of an annulus A_i whose boundary curves are J_{i1} and J_{i2}. However, for each $\epsilon > 0$, U_{01}, V_1^5, U_{04}, Q^7, V_2^5 could have been chosen to be ϵ-neighborhoods such that $A_0 - (U_{02} \cap A_0)$ is an ϵ-neighborhood in A_0 of the boundary curve of A_0 other than J_0. This proves the existence of the required sequence of annuli.

Corollary. The sequence $\{A_i\}$ may be so constructed that if J_{i1} and J_{i2} are the boundary curves of A_i, then the sequences $\{J_{i1}\}$ and $\{J_{i2}\}$ converge properly to the boundary curves of A_0.

Definition 4.11. A torus T_i in C_i'' will be called <u>unknotted</u> in C_i'' provided that it separates C_i'' and there are simple closed curves J_i' and J_i'' in T_i'' such that J_i' and J_i'' bound discs D_i' and D_i'' such that $D_i' - J_i'$ and $D_i'' - J_i''$ lie in different components of $C_i'' - T_i$.

Lemma 4.12. If T_0 is an unknotted torus in C_0'', then there is a sequence $\{T_i\}$ of unknotted tori converging strongly to T_0.

Proof. Let J_0' and J_0'' denote simple closed curves bounding unknotted annuli A_0' and A_0'' whose union is T_0. It follows from Theorem 4.10 that there is a sequence $\{A_i'\}$ of unknotted annuli converging to A_0' such that the boundary curves of A_i' may be labelled J_i' and J_i'' in such a way that the sequences $\{J_i'\}$ and $\{J_i''\}$ converge properly to J_0' and J_0''. Denote by W_0 and W_0' regular neighborhoods of A_0'' in E^n such that $W_0 \subset \text{int } W_0'$ and by x_0' and x_0'' simple closed curves in different components of

$(A_0' \cap W_0') - (A_0' \cap W_0)$ such that each separates J_0' from J_0'' in A_0'. It follows from Lemma 2.5* of [7] applied to the projection mapping of a regular neighborhood of A_0' onto A_0' that there are sequences $\{x_i'\}$ and $\{x_i''\}$ of simple closed curves converging strongly (in fact, properly) to x_0' and x_0'' such that for each i, x_i' and x_i'' are mutually exclusive sebsets of A_i' separating J_i' from J_i'' and bounding an annulus B_i in A_i'. An application of the proof of Lemma 3.4 of [7] demonstrates that the A_i' may be so selected that $\{B_i\}$ converges to B_0 and $\{A_i - B_i\}$ to a subset of W_0'. For sufficiently large i, $g_i'(J_i' \cup J_i'')$ and $g_i g_i'(J_i' \cup J_i'')$ lie in W_0. Also, $g_i'(J_i')$ and $g_i'(J_i'')$ bound a singular annulus in $W_0 \cap C_0''$. Thus $g_i g_i'(J_i')$ and $g_i g_i'(J_i'')$ bound a singular annulus in $W_0' - (W_0' \cap B_i)$ for sufficiently large i. But $g_i g_i'|J_i' \cup J_i''$ is homotopic to the identity map in $W_0' - (W_0' \cap B_i)$ for sufficiently large i, so that $J_i' \cup J_i''$ bounds a singular annulus in $W_0' - (W_0' \cap B_i)$. An application of the extension of Dehn's Lemma to discs with holes [18] implies that $x_i' \cup x_i''$ bounds an annulus E_i in $W_0' - (W_0' \cap \text{int } B_i)$ $\cup (A_i' - B_i)$. Thus $T_i = B_i \cup E_i$ is a torus and these may be constructed so that the sequence $\{T_i\}$ converges to T_0.

To see that this convergence is strong, consider a regular ε-neighborhood P_0 of T_0 in E^n and denote the projection mapping of P_0 into T_0 by p. Since the convergence of $\{x_i'\}$ and $\{x_i''\}$ to x_0' and x_0'' is proper and $\{B_i\}$ and $\{E_i\}$ converge to subsets of regular neighborhoods of A_i' and A_i'', it follows readily that $p_i|T_i$, where p_i is a piecewise linear ε_i-approximation to $p|T_i$, is homotopic to a homeomorphism of T_i into T_0 and thus takes a fundamental 2-cycle γ_i mod the integers carried by T_i onto a nonbounding cycle on T_0. Thus γ_i fails to bound in P_0 and the convergence of $\{T_i\}$ to T_0 is strong.

To see that T_i is unknotted for sufficiently large i, suppose that J_0 is a simple closed curve in T_0 carrying one of the generators of $H_1(T_0)$ and that J_0 bounds a disc D_0 which lies except for J_0 in V_0, one of the components of $C_0'' - T_0$. Denote the other component by U_0. Since the convergence of $\{T_i\}$ to T_0 is strong (and, in fact, proper), $C_i'' - T_i$ is the union of connected open sets U_i and V_i such that $\{U_i\}$ converges to $\text{cl}(U_0)$ and $\{V_i\}$ converges to $\text{cl}(V_0)$. The disc D_0 is a subset of a disc D_0' such that $D_0' \cap T_0 = J_0$ and $J_0' = \text{bdry } D_0' \subset U_0$. It may be supposed that P_0 does not contain J_0'. Also, let Q_0 be a regular neighborhood of J_0 in E^n which is consistently imbedded in P_0

and does not intersect J_0'. It follows from Lemmas 4.4 and 4.7 that there are a sequence $\{J_i'\}$ of simple closed curves converging properly to J_0' and a sequence $\{D_i'\}$ of discs converging to D_0' such that for each i, $J_i' = $ bdry D_i'. For sufficiently large i, $T_i \cap D_i' \subset Q_0$. Small adjustments may be made in D_i' so that each component of $D_i' \cap T_i$ is a simple closed curve. If each such component is contractible in Q_0, then J_0' bounds a disc in a regular neighborhood of $D_0' - $ int D_0, which contradicts the proper convergence of $\{D_i'\}$ to D_0'. Thus there is a disc E_i in D_i' whose boundary J_i is not contractible in Q_0 but which is such that every other component of $E_i \cap T_i$ is contractible in Q_0. If x_i is a component of $E_i \cap T_i$ which is contractible in Q_0, then x_i is contractible in P_0 and $p_i|x_i$ is homotopic to 0 on T_0. But, as was seen in the above paragraph, $p_i|T_i$ is homotopic to a homeomorphism of T_i onto T_0. Thus x_i is contractible in T_i and consequently bounds a disc in T_i. If F is a disc in T_i whose boundary lies in E_i but which is such that $E_i \cap$ int $F = 0$ and F' is the disc in E_i bounded by bdry F, replace F' by F and move it slightly away from T_i to obtain a disc E_i' bounded by J_i but having fewer simple closed curves in common with T_i than does E_i. If this process is repeated, a disc D_i is finally obtained which lies except for its boundary, J_i, in V_i. An application of this process to a disc in $U_0 \cup T_0$ completes the proof that T_i is unknotted for sufficiently large i.

Lemma 4.13. Suppose that T_0 is an unknotted torus in C_0'' and that $J_{01}, \ldots, J_{0m}$ are simple closed curves in T_0 such that for each j, $J_{0j} \cup J_{0j+1}$ (addition of subscripts being performed mod m) bounds an annulus A_{0j} in T_0, $\cup A_{0j} = T_0$ and $A_{0j} \cap A_{0k} = 0$ unless $|j - k| \leqq 1$. Then there is a sequence $\{T_i\}$ of tori converging properly to T_0 and for each i, there is a sequence $J_{i1}, \ldots, J_{im}$ of simple closed curves in T_i with the properties that for each j, $J_{ij} \cup J_{ij+1}$ bounds an annulus A_{ij} in T_i, $\cup_j A_{ij} = T_i$ and $A_{ij} \cap A_{ik} = 0$ unless $|j - k| \leqq 1$. Furthermore, the A_{ij} may be so selected that for each j, $\{A_{ij}\}$ converges strongly to A_{0j} and $\{J_{ij}\}$ converges strongly to J_{0j}.

Proof. This follows directly from an application of Lemma 4.12 and the proof of Lemma 3.5 of [7].

Lemma 4.14. Using the notation of Lemma 4.13, if $D_{01}, \ldots, D_{0m}$ are mutually exclusive discs bounded by $J_{01}, \ldots, J_{0m}$ such that for each j, $D_{0j} - J_{0j}$ lies in U_0 the component of $C_0'' - T_0$ which does not contain bdry C_0'', then the T_i may be so selected that for each i, j, there is a disc D_{ij} bounded by J_{ij} such that $D_{ij} - J_{ij} \subset U_i$, the component of $C_i{}'' - T_i$ which does not contain bdry $C_i{}''$, and $\{D_{ij}\}$ converges to D_{0j}.

Proof. In A_{0j}, let S_{0j} and R_{0j+1} denote mutually exclusive annuli, J_{0j} lying on bdry S_{0j} and J_{0j+1} lying on bdry R_{0j+1}. An application of Lemma 4.13 demonstrates that the annuli A_{ij} in T_i may be so selected that in A_{ij} there are mutually exclusive annuli S_{ij} and R_{ij+1}, J_{ij} lying on bdry S_{ij} and J_{ij+1} lying on bdry R_{ij+1}, and that the sequences of annuli $\{S_{ij}\}$, $\{R_{ij}\}$ and $\{cl[A_{ij} - (S_{ij} \cup R_{ij+1})]\}$ converge properly to S_{0j}, R_{0j} and $cl[A_{0j} - (S_{0j} \cup R_{0j+1})]$. If the part of the proof of Lemma 4.12 that demonstrates that T_i is unknotted is followed, it is seen that there is, in a regular neighborhood P_{0j} of D_{0j} which contains $S_{0j} \cup R_{0j}$, a disc D_{ij}' which is bounded by s simple closed curve J_{ij}' in $S_{ij} \cup R_{ij}$ such that the identity mapping of J_{ij}' onto itself is homotopic in $S_{ij} \cup R_{ij}$ to a homeomorphism onto J_{ij} and $D_{ij}' - J_{ij}' \subset U_i$. This follows from the fact that, using E_{ij} in place of E_i, components of $E_{ij} \cap T_i$ lie in $S_{ij} \cup R_{ij}$ for sufficiently large i. However, D_{ij}' may be moved very slightly away from T_i into U_i and J_{ij} and J_{ij}' may be joined by a singular annulus very close to $R_{ij} \cup S_{ij}$ which meets T_i only in points of J_{ij}. An application of Dehn's Lemma now yields a disc D_{ij} such that $J_{ij} = $ bdry D_{ij} and $D_{ij} - J_{ij} \subset U_i \cap P_{0j}$. Since $R_{0j} \cup S_{0j}$ and P_{0j} can be selected as ϵ-neighborhoods for each ϵ, the existence of the required sequences is demonstrated.

Lemma 4.15. Suppose that J_{00}, J_{01}, ..., J_{0m} are mutually exclusive simple closed curves in int $C_0{}''$, that D_{00} and D_{0m} are mutually exclusive discs bounded by J_{00} and J_{0m} respectively, that for each j, A_{0j} is an annulus bounded by $J_{0j-1} \cup J_{0j}$ such that $A_{0j} \cap A_{0j+1} = J_{0j}$ and $A_{0j} \cap A_{0k} = 0$ unless $|j - k| \leqq 1$ and that $D_{00} \cup D_{0m} \cup \bigcup A_{0j}$ is a 2-sphere S_0 in int $C_0{}''$. Then for each j there are a sequence $\{J_{ij}\}$ of simple closed curves converging 0-regularly to J_{0j} and a sequence $\{A_{ij}\}$ of annuli converging to A_{0j} such that bdry $A_{ij} = J_{ij-1} \cup J_{ij}$. The annuli A_{ij} may be so chosen that for each i and j, $A_{ij} \cap A_{ij+1} = J_{ij}$ and $A_{ij} \cap A_{ik} = 0$ unless $|j - k| \leqq 1$. Furthermore, there are sequences $\{D_{i0}\}$ and $\{D_{im}\}$ of discs converging to D_{00} and D_{0m} such that for each i, D_{i0} and D_{im} are bounded by J_{00} and J_{0m} and $D_{i0} \cup D_{im} \cup \bigcup A_{ij}$ is a 2-sphere S_i in int $C_i{}''$. It is clear that the sequence $\{S_i\}$ converges strongly to S_0.

Proof. Let ϵ be a positive number. Let R_{0j} denote, for $j = 0$, ..., m, an annulus which is a regular ϵ-neighborhood of J_{0j} in S_0 such that $R_{0j} \cap J_{0k} = 0$ unless $j = k$.

Denote by D_{00}' and D_{0m}' the discs $cl[D_{00} - (D_{00} \cap R_{00})]$ and $cl[D_{0m} - (D_{0m} \cap R_{0m})]$, by x_{0j} the boundary curve of R_{0j} in A_{0j} (or D_{00}, if $j = 0$), by z_{0j} the boundary curve of R_{0j} in A_{0j+1} (or D_{0m}, if $j = m$), and by S_{0j} the annulus $cl[A_{0j} - (A_{0j} \cap R_{0j-1}) - (A_{0j} \cap R_{0j})]$. For $j = 0, ..., m$, let T_{0j} denote a torus in C_0'' with interior U_{0j} such that $T_{0j} \cup U_{0j}$ is a regular ϵ-neighborhood of J_{0j} in C_0'' such that $S_0 \cap (U_{0j} \cup T_{0j})$ is an annulus in int R_{0j}. Suppose further, that V_{0j} is a regular ϵ-neighborhood of R_{0j} in E^n which contains $T_{0j} \cup U_{0j}$. For each j, there is a sequence $B_{0j1}, B_{0j2}, ..., B_{0jr}$ of annuli such that $\bigcup_k B_{0jk} = T_{0j}$, $B_{0jk} \cap B_{0jk+1}$ is a common boundary curve y_{0jk} of B_{0jk} and B_{0jk+1} and $B_{0jk} \cap B_{0jp} = 0$ unless $|k - p| \leq 1$ (addition of third subscripts being taken mod r), diam $B_{0jk} \leq \epsilon$ and y_{0jk} bounds a disc in $T_{0j} \cup U_{0j}$.

It follows from a slight and obvious modification of the proofs of Lemma 3 5 of [7] and Lemmas 4.12, 4.13, and 4.14 above that there are sequences $\{D_{i0}'\}$, $\{D_{im}'\}$, $\{R_{ij}\}$, $\{S_{ij}\}$, $\{x_{ij}\}$, and $\{z_{ij}\}$ converging properly to D_{00}', D_{0m}', R_{0j}, S_{0j}, x_{0j}, and z_{0j} such that for each i and j, D_{i0}' and D_{im}' are discs bounded by x_{i0} and z_{im}, R_{ij} is an annulus bounded by $x_{ij} \cup z_{ij}$, S_{ij} is an annulus bounded by z_{ij-1} and x_{ij}, and $D_{i0}' \cup D_{im}' \cup \bigcup(S_{ij} \cup R_{ij})$ is a 2-sphere S_i'. Clearly the sequence $\{S_i'\}$ converges strongly to S_0. (A 3-cell is added to the union of S_0 and its interior to make a solid torus to which Lemma 4.14 is applied.) Also, there are sequences $\{B_{ijk}\}_i$ and $\{y_{ijk}\}_i$ converging properly to B_{0jk} and y_{0jk} such that for each i, j, k, B_{ijk} is an annulus bounded by $y_{ijk-1} \cup y_{ijk}$ and $\bigcup_k B_{ijk}$ is a torus T_{ij}. The sequence $\{T_{ij}\}_i$ converges properly to T_{0j} for each j. Small adjustments may be made so that each component of $T_{ij} \cap S_i'$ is a simple closed curve.

For sufficiently large i, $R_{ij} \cup T_{ij} \subset$ int V_{0j}, $S_i' \cap T_{ij} \subset R_{ij}$ and separates x_{ij} from z_{ij} in R_{ij}, and $R_{ij} \cap T_{ij}$ lies in a small neighborhood of $R_{0j} \cap T_{0j}$ and intersects both of its components. It may be assumed that each component of $R_{0j} \cap T_{0j}$ is a simple closed curve which meets each y_{0jk} in a point and each B_{0jk} in an arc. Let R_{ij}' denote an annulus in R_{ij} one of whose boundary curves is x_{ij} and the other a component x_{ij}' of $R_{ij} \cap T_{ij}$ which is such that each other component of $R_{ij}' \cap T_{ij}$ bounds a disc in T_{ij}. Let R_{ij}'' denote an annulus in R_{ij} one of whose boundary curves is z_{ij} and the other a component z_{ij}'' of $R_{ij} \cap T_{ij}$ which is such that each other component of $R_{ij}'' \cap T_{ij}$ bounds a disc in T_{ij}. (If a component x of $R_{ij} \cap T_{ij}$ bounds a disc in T_{ij},

x and x_{ij} do not bound an annulus in R_{ij} for if they do, each 1-cycle carried by x_{ij} bounds in V_{ij}. Hence x bounds a disc in R_{ij}.) If E is a disc in T_{ij} such that bdry $E = (R_{ij}' \cup R_{ij}'') \cap E$, then replace the disc in $R_{ij}' \cup R_{ij}''$ bounded by bdry E by E and move it slightly away from T_{ij}. Repeat this process until annuli R_{ij}^x and R_{ij}^z are obtained in $V_{ij} \cap C_i''$ such that bdry $R_{ij}^x = x_{ij} \cup x_{ij}'$, bdry $R_{ij}^z = z_{ij} \cup z_{ij}''$, $R_{ij}^x \cap R_{ij}^z = 0$ and $(R_{ij}^x \cup R_{ij}^z) \cap T_{ij} = x_{ij}' \cup z_{ij}''$. Note that this process does not alter S_i' - int R_{ij}.

Denote by γ_{ij}' and γ_{ij}'' fundamental generating 1-cycles of $H_1(x_{ij}')$ and $H_1(z_{ij}'')$, by a_{ij} a generating 1-cycle of $H_1(J_{ij}'')$, where J_{ij}'' is a simple closed curve in T_{ij} which meets each B_{ijk} in an arc and each y_{ijk} in a point and by β_{ij} a generating 1-cycle of $H_1(y_{ijk})$. Then $\gamma_{ij}^* \sim a_{ij}^* a_{ij} + b_{ij}^* \beta_{ij}$ in T_{ij} where (*) denotes (') or ('') and a_{ij}^* and b_{ij}^* are integers. Thus if W_{0j} denotes a regular ϵ-neighborhood in E^n of T_{0j} and p_{ij} is a piecewise linear ϵ_i-approximation to the restriction to T_{ij} of the projection map of W_{0j} onto T_{0j}, then $p_{ij}(\gamma_{ij}^*) \sim a_{ij}^* p_{ij}(a_{ij}) + b_{ij}^* p_{ij}(\beta_{ij})$ in T_{0j}. The proper convergence implies that if η_{0j} denotes a suitably oriented 1-cycle generating the first homology group of a component of $R_{0j} \cap T_{0j}$, then $\eta_{0j} \sim p_{ij}(\gamma_{ij}^*)$ in T_{0j} for suitably oriented γ_{ij}' and γ_{ij}'' and $p_{ij}(\gamma_{ij}^*) \sim a_{ij}^*(a_{0ij}) + b_{ij}^*(\beta_{0j})$, where a_{0ij} is a generator of first homology group of a simple closed curve in T_{0j} which meets each B_{0jk} in an arc and each y_{0jk} in a point and the orientations of a_{0ij} and β_{0j} depend on i. It follows that $a_{ij}' = a_{ij}'' = \pm 1$ and $b_{ij}' = b_{ij}''$ and that the carrier of a_{ij} could have been chosen so that $b_{ij}' = b_{ij}'' = 0$.

Therefore the identity map of x_{ij}' onto itself is homotopic in T_{ij} to a homeomorphism of x_{ij}' onto z_{ij}'' and to a homeomorphism onto a simple closed curve J_{ij}' which meets each B_{ijk} in an arc and each y_{ijk} in a point. There is a homeomorphism f_i of $T_{ij} \times I$ into $U_{ij} \cup T_{ij}$, where U_{ij} is the component of $C_i'' - T_{ij}$ that lies in V_{0j}, carrying $T_{ij} \times (0)$ onto T_{ij} and $T_{ij} \times (1)$ onto a torus T_{ij}' such that $d(f_i(x,t), f_i(x,t')) < \epsilon/100$. Let J_{ij} denote $f_i(J_{ij}', 1)$. The remarks above concerning homotopy of the identity map of x_{ij}' imply that J_{ij} and x_{ij}' bound a singular annulus in $f_i(T_{ij} \times I)$ and, as a consequence of the extension of Dehn's Lemma to discs with holes [18], a non-singular annulus S_{ij}^x in $f_i(T_{ij} \times I)$ such that $S_{ij}^x \cap (T_{ij} \cup T_{ij}') = x_{ij}' \cup J_{ij}$. Also, J_{ij} and z_{ij}'' bound a non-singular annulus S_{ij}'' in $f_i(T_{ij} \times I)$ such that $S_{ij}'' \cap (T_{ij} \cup T_{ij}') = z_{ij}'' \cup J_{ij}$ and each component of $S_{ij}^x \cap S_{ij}''$ is a simple closed curve at which the two annuli actually cross each other (except for J_{ij}). The usual procedure for removing components implies that it can be assumed that no component of $S_{ij}{}^X \cap S_{ij}''$ bounds a disc in $S_{ij}{}^X$ or S_{ij}''. If E is the annulus in S_{ij}'' such that $E \cap S_{ij}{}^X$ is a boundary curve of E, then $cl(S_{ij}'' - E)$ may be replaced by the annulus in $S_{ij}{}^X$ bounded by J_{ij} and $E \cap S_{ij}{}^X$ and then moved slightly away from $S_{ij}{}^X$ to yield an annulus $S_{ij}{}^Z$ which lies in $f_i(T_{ij} \times I)$ but meets $S_{ij}{}^X$ only in J_{ij}.

The annulus $A_{ij} = S_{ij-1}{}^Z \cup R_{ij-1}{}^Z \cup S_{ij} \cup R_{ij}{}^X \cup S_{ij}{}^X$ is bounded by $J_{ij-1} \cup J_{ij}$ for $j = 1, \ldots, m$, the discs $D_{i0} = D_{i0}' \cup R_{i0}{}^X \cup S_{i0}{}^X$ and $D_{im} = D_{im}' \cup R_{im}{}^Z \cup S_{im}{}^Z$ are bounded by J_{i0} and J_{im} and $D_{i0} \cup D_{im} \cup \bigcup A_{ij}$ is a 2-sphere S_i. Since this construction may be made for each ϵ and each selection of T_{0j}, the existence of the required sequences is proved.

<u>Theorem 4.16.</u> If S_0 is a 2-sphere in int C_0'', then there is a sequence $\{S_i\}$ of 2-spheres converging h - 1-regularly to S_0.

<u>Proof.</u> The proof follows that of Theorem 3.10 of [7], which is a slightly modified form of 4.16 for the case where each M_i is polyhedrally imbeddable in E^3, without essential change, so only an outline will be given here. Denote by $s_{00}, s_{01}, \ldots, s_{0m}$ a sequence of mutually exclusive simple closed curves in S_0 such that for each $j = 1, \ldots, m$, $s_{0j-1} \cup s_{0j}$ bounds an annulus A_{0j} in S_0, $A_{0j} \cap A_{0j+1} = s_{0j}$, $A_{0j} \cap A_{0k} = 0$ unless $|j - k| \leqq 1$ and s_{00} and s_{0m} bound mutually exclusive discs D_{00} and D_{0m} in S_0. Denote by $t_{00}, t_{01}, \ldots, t_{0m}$ a sequence of mutually exclusive arcs in S_0 such that for each j and k, $t_{0k} \cap s_{0j}$ is a point P_{jk}, $t_{0k-1} \cup t_{0k}$ lies on the boundary of a component of $S_0 - \bigcup t_{0j} - s_{00} - s_{0m}$ which is a disc B_{0k}, $B_{0k} \cap A_{0j}$ is a disc D_{0jk} and $t_{0k} \cap s_{00}$ and $t_{0k} \cap s_{0m}$ are the endpoints of t_{0k}. It follows from Lemma 4.15 that there are sequences $\{s_{ij}\}$, $\{A_{ij}'\}$, $\{S_i'\}$, $\{D_{i0}\}$ and $\{D_{im}\}$ converging regularly to s_{0j} and strongly to A_{0j}, S_0, D_{00} and D_{0m} such that for each i and j, $s_{ij-1} \cup s_{ij}$ is the boundary of the annulus A_{ij}', $A_{ij}' \cap A_{ij+1}' = s_{ij}$, $A_{ij}' \cap A_{ik}' = 0$ unless $|j - k| \leqq 1$, s_{00} and s_{0m} are the boundaries of the discs D_{i0} and D_{im} and $D_{i0} \cup D_{im} \cup \bigcup A_{ij}'$ is the 2-sphere S_i'.

An application of Lemma 2.5* of [7] and the regular convergence of $\{s_{ij}\}$ to s_{0j} demonstrate the existence for each k of a sequence $\{t_{ik}\}$ of arcs converging properly to t_{0k} such that $t_{ik} \cap s_{ij}$ is a point P_{ijk}, $t_{ik} \subset S_i'$ and $t_{ik} \cap s_{i0}$ and $t_{ik} \cap s_{im}$ are the endpoints of t_{ik}. Since the number m is arbitrary and the A_{0j}'s could be quite narrow, it may be assumed that the convergence of $\{t_{ik}\}$ to t_{0k} is regular. The set $t_{ik-1} \cup t_{ik}$

lies on the boundary of a disc B_{ik}' which is the closure of a component of $S_i' - \bigcup t_{ij}$
$- s_{i0} - s_{im}$ and $A_{ij}' \quad B_{ik}'$ is a disc D_{ijk}'. The sequence of discs $\{D_{ijk}'\}_i$ may not
converge to D_{0jk} but it does converge to a subset of A_{0j}. The proof of Theorem 3.10 of
[7] now demonstrates the existence of an integer p independent of m and the s_{0j} and t_{0j}
with the property that the D_{ijk}' may be adjusted to give discs D_{ijk} such that
$D_{i0} \cup D_{im} \cup \bigcup_{jk} D_{ijk} = S_i$, a 2-sphere, and $\{D_{ijk}\}$ converges to a connected subset of
S_0 which contains D_{0jk} and is contained in the union of p of the discs D_{00}, D_{0m}, D_{0jk}.
But for each positive number ϵ, there are sequences $\{s_{0j}\}$ and $\{t_{0j}\}$ for which the
connected union of p of the discs D_{0jk} has diameter less than ϵ. Thus it is seen that
the required sequence of 2-spheres exists.

<u>Corollary.</u> The convergence of $\{S_i\}$ to S_0 is completely regular.

<u>Proof.</u> This is a direct consequence of the h - 1-regularity of the convergence.
See Lemma 3 of [6].

<u>Lemma 4.17.</u> Suppose that S_0 is a 2-sphere in M_0 bounding a 3-cell C_0 such that
$S_0 \cap K_0$ is a compact 2-manifold with boundary, D_0, and that there is a sequence $\{S_i\}$
of 2-spheres converging completely regularly to S_0 such that for each i, $S_i \cap K_i = g_i(D_0)$.
Denote by C_i the closure of the component of $M_i - S_i$ whose closure contains $g_i(D_0)$.
Then the sequences $\{C_i\}$ and $\{cl(M_i - C_i)\}$ converge h - 2-regularly to C_0 and
$cl(M_0 - C_0)$.

<u>Proof.</u> This is essentially the second part of Lemma 4.3 of [7] and could be
proved in a fashion similar to the proof of Lemma 3.11 of [7] but a simpler argument is
given here.

That the C_i and $M_i - C_i$ exist and converge to C_0 and $cl(M_0 - C_0)$ was demonstrated in the proof of Lemma 4.1. Let ϵ be a positive number. Suppose that U_0 is a
regular $\epsilon/8$-neighborhood of S_0 in M_0. Let δ be a positive number such that each mapping of a k-sphere, $k \leq 2$, into a subset of M_i of diameter less than δ is homotopic to
0 in a subset of M_i of diameter less than $\epsilon/8$ and let f_i be a mapping of the k-sphere
J bounding a (k + 1)-cell C into a subset of C_i of diameter less than δ. Then f_i may be
extended to a mapping F_i of C into M_i such that diam $F_i(C) < \epsilon/8$. It may be assumed
that F_i is piecewise linear.

It follows from the construction of the C_i and from Lemma 2.6 of [7] that for sufficiently large i there are piecewise linear $\epsilon/8$-mappings h_i and h_i' such that (1) h_i' maps $S_i \cup [F_i(C) \cap cl(M_i - C_i)]$ into U_0, (2) $h_i'|S_i$ is a homeomorphism onto S_C, (3) h_i is a homeomorphism of S_0 onto S_i and (4) $h_i'|S_i = h_i^{-1}$. Also, there is an $\epsilon/4$-mapping k_i carrying U_0 onto S_0 which leaves each point of S_0 fixed. Denote by G_i the mapping of C into C_i such that $G_i(x) = F_i(x)$ if $F_i(x) \in C_i$ and $G_i(x) = h_i k_i h_i' F_i(x)$ if $F_i(x) \in M_i - C_i$. Clearly diam $G_i(C) < \epsilon$ and the lemma is proved.

<u>Corollary.</u> The sets C_i are homotopy cells in the sense that their homotopy groups all vanish and they are compact 3-manifolds with a 2-sphere for boundary.

<u>Proof.</u> Since the convergence of $\{C_i\}$ to C_0 is completely regular, it follows from Lemma 2.6* of [7] that there are sequences $\{h_i\}$ and h_i' of mappings such that for each i, h_i maps C_0 piecewise linearly onto C_i, h_i' maps C_i piecewise linearly onto C_0, $h_i|S_0$ and $h_i'|S_i$ are inverse homeomorphisms and neither moves any point as much as ϵ_i. Suppose that f_i maps J into C_i. Then $h_i'f_i$ maps J into C_0 and can be extended to a mapping F_i of C into C_0. Then $h_i F_i$ is a mapping of C into C_i such that if $x \in J$ then $d(f_i(x), h_i F_i(x)) < 2\epsilon_i$. Since $\{\epsilon_i\}$ converges to 0, the regularity of the convergence of $\{C_i\}$ to C_0 implies that for sufficiently large i, there is a mapping k_i of $J \times I$ into C_i such that $k_i(x,0) = f_i(x)$ and $k_i(x,1) = h_i F_i(x)$. A combination of the homotopies effected by k_i and $h_i F_i$ yields an extension of f_i to a mapping of C into C_i.

<u>Note.</u> It is known that if each homotopy cell is a 3-cell, then the Poincaré conjecture is true and conversely.

<u>Lemma 4.18.</u> If S_0 is a disc or annulus in M_0 and $S_0 \cap K_0 = $ bdry S_0, then there is a sequence $\{S_i\}$ of discs (annuli) converging completely regularly to S_0 such that for each i, $S_i \cap K_i = g_i(S_0 \cap K_0)$.

<u>Proof.</u> This Lemma is essentially Lemma 4.1 of [7]. Its proof is a repetition of the proofs of Theorem 4.16 and the lemmas and theorems preceding it. The case where $S_0 \cap K_0$ is not connected is treated first since S_0 then has properties similar to those of the tori of Lemmas 4.12, 4.13, and 4.14. The connected case may be treated by applying first a modification of the proof of Lemma 4.15.

<u>Lemma 4.19.</u> If S_0 is a 2-sphere bounding a 3-cell C_0 in M_0 and $S_0 \cap K_0$ is a

compact 2-manifold with boundary, then there is a sequence $\{S_i\}$ of 2-spheres converging completely regularly to S_0 such that for each i, $S_i \cap K_i = g_i(S_0 \cap K_0)$.

Proof. The lemma is essentially Lemma 4.3 of [7] and the proof of that Lemma may be applied here. The proof follows from Lemma 4.17 and an induction argument on the number of components of $S_0 \cap K_0$.

Corollary. If for each i, C_i denotes the closure of the component of $M_i - S_i$ whose closure contains $S_i \cap K_i$, then the sequences $\{C_i\}$ and $\{cl(M_i - C_i)\}$ converge h - 2-regularly to C_0 and $cl(M_0 - C_0)$.

Proof. This follows from Lemma 4.1 and Lemma 4.17.

Definition 4.20. If M is a 3-manifold with boundary K, which may be empty, then a cellular decomposition of M is a finite collection G of polyhedral 3-cells whose union is M and which is such that the intersection, if it exists, of each element of G with the union of any number of elements of the collection consisting of K and the remaining elements of G is a compact 2-manifold with boundary.

Such a decomposition is described by Bing in [3], p. 17. If x is an element of G, then M - x is a compact 3-manifold with boundary, perhaps not connected, and the decomposition of M - int x consisting of the elements of G other than x is a cellular decomposition.

Definition 4.21. The collection G is a pseudo-cellular decomposition of M if each of its elements is a homotopy 3-cell and it otherwise satisfies the conditions of Definition 4.20.

Theorem 4.22. If M_0^2 is the 2-skeleton of a cellular decomposition G_0 of M_0 and ϵ is a positive number, then there is, for sufficiently large i, an ϵ-homeomorphism h_i of M_0^2 into M_i which extends g_i and is such that $h_i(M_0^2)$ is the 2-skeleton M_i^2 of a pseudo-cellular decomposition G_i of M_i.

Proof. If the cellular decomposition G_0 consists of only one element, then the theorem is obviously true. Suppose the theorem to be true for each 3-manifold with boundary that has a cellular decomposition with fewer than k elements and each sequence of 3-manifolds converging h - 2-regularly to it as in the principal hypotheses of this paper and that G_0 is a cellular decomposition of M_0 with k elements. Let x_0 denote an element of G_0. It follows from Theorem 4.16, Lemma 4.19 and their corollaries that there are a sequence $\{\delta_i\}$ of positive numbers converging to 0 and a sequence $\{h_{ix}\}$ of homeomorphisms such that for each i, h_{ix} is a δ_i-homeomorphism of $K_0 \cup$ bdry x_0 into M_0 which extends g_i and is such that the closure, x_i, of one of the components of $M_i - h_{ix}($ bdry $x_0)$ is a homotopy 3-cell with the property that $\{x_i\}$ converges h - 2-regularly to x_0 and $\{cl(M_i - x_i)\}$ converges h - 2-regularly to $cl(M_0 - x_0)$. Each $cl(M_i - x_i)$ is a compact 3-manifold with boundary and $cl(M_0 - x_0)$ has a cellular decomposition with fewer than k-elements. The induction hypothesis now yields the theorem.

Theorem 4.23. If the Poincaré conjecture is true or if each homotopy cell in each M_i is a 3-cell, then the convergence of $\{M_i\}$ to M_0 is completely regular.

Proof. Suppose that ϵ is a positive number and that a_0 is a cellular decomposition of M_0 each cell of which has diameter less than $\epsilon/6$. Denote the 2-skeleton of a_0 by A^2. For sufficiently large i, there is an $\epsilon/6$-homeomorphism h_i of A^2 into M_i which extends g_i and is such that if x_0 is a 3-cell in a_0, then $h_i($ bdry $x_i)$ bounds a homotopy cell x_i in M_i which lies in an $\epsilon/6$-neighborhood of x_0. But x_i is a 3-cell and $h_i |$ bdry x_0 may be extended to a homeomorphism of x_0 onto x_i which is an ϵ-homeomorphism. Thus h_i may be extended to an ϵ-homeomorphism of M_0 onto M_i and the theorem is proved.

An immediate consequence is

Theorem 4.24. If f is an h - 2-regular mapping of a metric space X onto a metric space Y such that each inverse under f is a compact 3-manifold with boundary and there is a compact 2-manifold to which each bdry $f^{-1}(y)$ is homeomorphic for y in Y, then if (1) the Poincaré conjecture is true or (2) each homotopy cell in each $f^{-1}(y)$ is a 3-cell, then f is completely regular.

5. <u>The space of homeomorphisms on a 3-manifold.</u> In our paper [6], Eldon Dyer and I proved in Theorem 1 that the space of homeomorphisms of a 2-manifold with boundary which leave the boundary pointwise fixed is locally contractible. This result contains a result of Kneser's and uses his techniques. Fort and Roberts also published, independently, proofs of parts of Kneser's results. In this section it is proved (Theorem 5.2) that the space of homeomorphisms of a 3-manifold with boundary onto itself

which leave the boundary pointwise fixed in LC^n for each n. (A space X is LC^n if for each point x in X and each positive number ϵ there is a positive number δ such that each mapping of a k-sphere, $k \leqq n$, into $S(x,\delta)$ is homotopic to 0 in $S(x,\epsilon)$.) A well known theorem of Alexander [1] asserts that the space of homeomorphisms of an n-cell into itself leaving its boundary pointwise fixed is locally contractible. The reader is also referred to the work of Fisher, Kister and Sanderson mentioned in the introduction.

Use will be made of a selection theorem of E. Michael [10], which is stated here in weaker form than that in which Michael states it.

Theorem M. Suppose that A and B are metric spaces, A complete and the (covering) dimension of B does not exceed n + 1, and that Z is a closed subset of B. Suppose, further, that Q is an open mapping of A onto B such that the collection of inverses under Q is equi-LC^n (see below) and q is a mapping of Z into A such that for $z \in Z$, $q(z) \in Q^{-1}(z)$. Then there is a neighborhood U of Z such that q may be extended to a mapping q* of U into A such that for $z \in U$, $q*(z) \in Q^{-1}(z)$. If each inverse under Q has the property that all its homotopy groups vanish (of order $< n + 1$), then U may be taken to be the entire space B.

The collection of inverses under Q is said to be equi-LC^n provided that if $b \in B$, $a \in Q^{-1}(b)$ and $\epsilon > 0$, then there is a $\delta > 0$ such that if $b' \in B$, then every mapping of a k-sphere, $k \leqq n$, into $Q^{-1}(b') \cap S(a,\delta)$ is homotopic to 0 in $Q^{-1}(b') \cap S(a,\epsilon)$.

If f and g are elements of the space of homeomorphisms of a compact metric space X into a compact metric space Y, then the distance between f and g is defined to be $d(f,g) = \text{lub } d(f(x), g(x))$ for $x \in X$.

Suppose that C_1 is a 3-cell and C_2 is a 3-cell in C_1 such that $C_2 \cap$ bdry C_1 is a 2-manifold with boundary. Denote by H_1 the space of homeomorphisms of C_1 onto itself leaving bdry C_1 pointwise fixed, by i_1 its identity and by $S_1(i_1, \epsilon)$ the (open) ϵ-neighborhood of i_1 in H_1. Denote by H_2 the space of homeomorphisms of C_2 into C_1 leaving $C_2 \cap$ bdry C_1 pointwise fixed and carrying $C_2 - (C_2 \cap$ bdry $C_1)$ into int C_1, by i_2 its identity and by $S_2(i_2, \epsilon)$ the (open) ϵ-neighborhood of i_2 in H_2. Denote by K the closure of $C_1 - C_2$, by H the space of homeomorphisms of K into itself leaving bdry K pointwise fixed, by i its identity and by $S(i, \epsilon)$ the ϵ-neighborhood of i in H. The space H_2

should be further restricted to consist only of those homeomorphisms f for which $cl(C_1 - f(C_2))$ is a 3-manifold with boundary.

Theorem 5.1. If n is an integer and ϵ is a positive number, then there is a positive number δ such that (1) if f is a mapping of R^n into $S_2(i_2, \delta)$, then there is a mapping F of R^n into $S_1(i_1, \epsilon)$ such that for each x in R^n, $F(x)|C_2 = f(x)$ and (2) if f is a mapping of S^n into $S_2(i_2, \delta)$, then there is a mapping F of S^n into $S_1(i_1, \epsilon)$ such that for each x in S^n, $F(x)|C_2 = f(x)$.

Theorem 5.2. If M is a compact 3-manifold with boundary, then the space of homeomorphisms of M onto itself leaving the boundary of M pointwise fixed is LC^n for each n.

Proof of Theorems 5.1 and 5.2. The techniques of section 4 and of [7] may be used to prove that there is a positive number δ such that each element of $S_2(i_2, \delta)$ may be extended to an element of $S_1(i_1, \epsilon)$. Also, Lemma 5.3 of [7], which follows from Moise's Lemma on the fitting together of homeomorphisms [12], is a statement of a special case of Theorem 5.1 for the case n = 0. Lemma 5.4 of [7] states that the space of homeomorphisms of M onto itself leaving bdry M pointwise fixed is LC^0. The proof will follow by induction on n. Assume that Theorems 5.1 and 5.2 are true for $n \le k$. It will be proved first that part (1) of Theorem 5.1 is true for n = k + 1.

Let ϵ be a positive number. It follows from the induction hypothesis that H is LC^k and from Lemma A of [5] that the space $H(R^k, K)$ of homeomorphisms of $R^k \times K$ onto itself which leave each (x,K) invariant and $R^k \times$ bdry K pointwise fixed is locally connected. Thus there is a positive number $\delta_1 < \epsilon/4$ such that if f is a $2\delta_1$-homeomorphism in $H(R^k, K)$, then there is a mapping F of I into $H(R^k, K)$ such that F(0) is the identity, F(1) = f and f(t) is an $\epsilon/4$-homeomorphism for each t. The induction hypothesis implies that there is a positive number δ such that if f is a mapping of R^k into $S_2(i_2, \delta)$, then there is a mapping F of R^k into $S_1(i_1, \delta_1)$ such that for each x in R^k, $F(x)|C_2 = f(x)$.

Consider the space $R^{k+1} \times C_1$ with its usual metric and the mapping f of R^{k+1} into $S_2(i_2, \delta)$. Denote by f^* the homeomorphism of $R^{k+1} \times (C_2 \cup$ bdry $C_1)$, considered as a subset of $R^{k+1} \times C_1$, into $R^{k+1} \times C_1$ which is such that $f^*(x,c) = (x, f(x)(c))$ for each c in C_2 and $f^*(x,c) = (x,c)$ for c in bdry C_1. Note that, in the usual metric for $R^{k+1} \times C_1$, f^* is a δ-homeomorphism and that $d[(x,c), f^*(x,c)] = d[x, f(x)(x)]$. The principal part of the following argument is the establishment of an extension F^* of f^* to an ϵ-homeo-morphism of R^{k+1} onto itself which leaves each (x, C_1) invariant.

Consider R^{k+1} as the product $R^k \times I$. The induction hypothesis asserts that for t in I there is a mapping f_t of (R^k, t) into $S_1(i_1, \delta_1)$ such that $f_t(x,t)|C_2 = f(x,t)$ for each (x,t) in $R^k \times I$. Denote by f_t^* the homeomorphism of $R^k \times t \times C_1$ onto itself which is such that $f_t^*(x,t,c) = (x,t,f_t(x,t)(c))$ and by g_t the homeomorphism of $R^k \times K$ into $R^k \times t \times C_1$ which is such that $g_t(x,y) = (x,t,f_t(x,t)(y))$. Note that f_t^* is an extension of $f^*|R^k \times t \times C_2$ to $R^k \times t \times C_1$ with the required properties and that $d[(x,t,y), g(x,y)]$ $< \delta_1$. Let $S(t)$ denote the space of homeomorphisms of $R^k \times K$ onto $g_t(R^k \times K)$ which extend $g_t|R^k \times$ bdry K and which is such that if $z \in S(t)$, then $z(x,K) = g_t(x,K)$ and $d[(x,t,y), z(x,y)] < \delta_1$. Denote by p a mapping of $U_{g_t}(R^k \times K)$ onto I which carries $g_t(R^k \times K)$ onto t for each t in I. It is seen from the construction that p is continuous. Also, $f^*|R^k \times I \times$ bdry K is a homeomorphism onto $U_{g_t}(R^k \times$ bdry $K)$ which carries $(x,t,$bdry $K)$ onto $g_t(x,$bdry $K)$. Some Lemmas will provide needed properties of $S(t)$ and $U S(t)$, considered as a subspace of the space of all homeomorphisms of $R^k \times K$ into $U_{g_t}(R \times K)$. For similar theorems, the reader is referred to [5], particularly to the proof of Theorem 5, which is quite similar in most details.

<u>Lemma 5.3.</u> The mapping p is completely regular in the strong sense that if ϵ is a positive number, then there is a positive number δ such that if $t \in I$ and $|t - t'| < \delta$, then there is an ϵ-homeomorphism z of $g_t(R^k \times K)$ onto $g_{t'}(R^k \times K)$ which carries each set $g_t(x,K)$ onto $g_{t'}(x,K)$ and is such that $zg_t|R^k \times$ bdry $K = g_{t'}|R^k \times$ bdry K.

<u>Proof.</u> First consider the case $k = 0$. Then $R^{k+1} = I$. It follows from the construction of g_t and a modification of Lemma 4.17 that p is h - 2-regular. Thus it follows from Theorem 4.5 of [7] that p is completely regular in the ordinary sense. If ϵ is a positive number, then it follows from Lemma 5.2 of [7] that there is a positive number $\delta' < \epsilon$ such that each $\delta'/2$-homeomorphism of g_t (bdry K) onto itself is extend-able to an $\epsilon/2$-homeomorphism of $g_t(K)$ onto itself and there is a positive number δ such that if $|t - t'| < \delta$, then (1) there is a $\delta'/4$-homeomorphism h of $g_t(K)$ onto $g_{t'}(K)$ and (2) $g_{t'}g_t^{-1}|g_t($bdry $K)$ moves no point as much as $\delta'/4$. (Statement (2) is a conse-quence of the construction of the maps g_t.) The homeomorphism $h^{-1}g_{t'}g_t^{-1}$ of $g_t($bdry $K)$ onto itself moves no point as much as $\delta'/2$ and thus may be extended to an $\epsilon/2$-homeo-morphism H of $g_t(K)$ onto itself. Then $z = hH$ is an ϵ-homeomorphism of $g_t(K)$ onto

$g_t(K)$ onto $g_{t'}(K)$ and $zg_t|\text{bdry } K \equiv g_{t'}|\text{bdry } K$.

If k is positive, then the argument for Lemma C of [5] will prove Lemma 5.3.

Lemma 5.4. The space $\bigcup S(t)$ may be remetrized so as to be complete.

Proof. The proof is essentially a repetition of the proof of Lemma 2 of [5].

Lemma 5.5. The collection of all $S(t)$ is equi-LC^0 and lower semi-continuous (i.e. the mapping Q of $\bigcup S(t)$ into I carrying each $S(t)$ into t is open).

Proof. This is essentially a restatement of the proofs of Lemmas 3 and 4 of [5].

Continuation of the proof of Theorems 5.1 and 5.2. The Lemmas and Theorem M now demonstrate that there is a sequence $0 = a_0 < a_1 < \ldots < a_m = 1$ such that for each i, there is a mapping q_i of the interval $[a_{i-1}, a_i]$ into $\bigcup S(t)$ such that for $a_{i-1} \lesseqgtr t \lesseqgtr a_i$, $q_i(t) \in S(t)$. The mapping Q_i of $R^k \times [a_{i-1}, a_i] \times K$ onto $\bigcup g_t(R^k \times K)$, $t \in [a_{i+1}, a_i]$, which carries the point (x, t, y) onto $q_i(t)(x, y)$ is a homeomorphism, $d[(x, t, y), q_i(t)(x, y)] < \delta_1$ and Q_i extends $f^*|R^k \times [a_{i-1}, a_i] \times \text{bdry } K$.

The homeomorphisms Q_i and Q_{i+1} may not coincide on $R^k \times a_i \times K$. Consider the homeomorphisms $q_i(a_i)$ and $q_{i+1}(a_i)$ of $R^k \times K$ onto $g_{a_i}(R^k \times K)$ and the homeomorphisms $p_i = [q_i(a_i)]^{-1}q_{i+1}(a_i)$ of $R^k \times K$ onto itself. If $p_i(x, y) = (x', y')$, it is seen that $d[(x, y), (x', y')] = d[(x, a_i, y), (x', a_i, y')] \lesseqgtr d[(x, a_i, y), q_{i+1}(a_i)(x,)] + d[q_{i+1}(a_i)(x, y), (x', a_i, y')] = d[(x, a_i, y), q_{i+1}(a_i)(x, y)] + d[q_i(a_i)(x', y'), (x', a_i, y')] < 2\delta_1$. Thus there is a mapping z_i of I into $H(R^k, K)$ such that $z_i(0)$ is the identity, $z_i(1) = p_i$ and each $z_i(t)$ is an $\epsilon/4$-homeomorphism.

The mapping $q_i(t)z_i[(t - a_{i-1})/(a_i - a_{i-1})] = Z_i(t)$, $a_{i-1} \lesseqgtr t \lesseqgtr a_i$ is a homeomorphism of $R^k \times K$ into $g_t(R^k \times K)$, $Z_i(a_{i-1}) = q_i(a_{i-1})$ and $Z_i(a_i) = q_i(a_i)p_i = q_{i+1}(a_i)$. Furthermore, $Z_i(t)|R^k \times \text{bdry } K = g_t|R^k \times \text{bdry } K$ and $d[(x, t, y), Z_i(t)(x, y)] < \epsilon/2$. Denote by F^* the $\epsilon/2$-homeomorphism of $R^k \times I \times C_1$ onto itself which is such that $F^*|R^k \times I \times (C_2 \cup \text{bdry } C_1) = f^*$ and $F^*(x, t, y) = Z_i(t)(x, y)$ for $y \in K$ and $a_{i-1} \lesseqgtr t \lesseqgtr a_i$.

Let F be the mapping of $R^{k+1} = R^k \times I$ into H_1 which is such that $F^*(x, t, y) = (x, t, F(x, t))$. It follows from the construction that $F(x, t)$ is an c-homeomorphism and that $F(x, t)|C_2 = f(x, t)$. Thus part (1) of Theorem 5.1 is proved for $n = k + 1$.

Lemma 5.6. The space $H(S^k, K)$ of homeomorphisms of $S^k \times K$ onto itself that leave each (x,K) invariant and leave $S^k \times$ bdry K pointwise fixed is locally connected.

$\underline{Proof.}$ Since $H(S^k,K)$ may be given a group structure, it need only be proved that it is LC^0 at the identity. Let ϵ be a positive number and δ a positive number, which, by the induction hypothesis, exists, such that each mapping of S^k into $S(i,\delta)$ is homotopic in $S(i,\epsilon)$ to a map of S^k onto i and let f be a δ-homeomorphism in $H(S^k,K)$. For each x in S^k, let $g(x)$ denote the homeomorphism of K onto itself which is such that $f(x,y) = (x,g(x)(y))$. Then g is a mapping of S^k into $S(i,\delta)$. Thus there is a mapping g^* of $S^k \times I$ into $S(i,\epsilon)$ such that $g^*(x,1) = g(x)$ and $g^*(x,0) = i$ for each x in S^k. Denote by $F^*(t)$ the mapping of $S^k \times K$ into itself which is such that $F^*(t)(x,y) = (x,g^*(x,t(y))$. Then $F^*(t)$ is an ϵ-homeomorphism, F^* is a continuous mapping of I, $F^*(t)$ leaves (x,K) invariant and $(x,$ bdry $K)$ pointwise fixed, $F^*(1)(x,y) = (x,g^*(x,1)(y)) = (x,g(x)(y)) = f(x,y)$, and $F^*(0)(x,y) = (x,g^*(x,0)(y)) = (x,y)$. The local connectivity is now established.

$\underline{Proof\ of\ part\ (2)\ of\ Theorem\ 5.1\ for\ the\ case\ n = k + 1.}$ Suppose that ϵ is a positive number. There is a positive number $\delta_1 < \epsilon/4$ such that if $f \in H(S^k,K)$ and moves no point as much as $2\delta_1$, then there is a mapping q of I into $H(S^k,K)$ such that $q(1) = f$, $q(0)$ is the identity and each $q(t)$ is an $\epsilon/4$-homeomorphism. There is a positive number δ such that if f is a mapping of R^{k+1} into $S_2(i_2,\delta)$, then there is a mapping F of R^{k+1} into $S_1(i_1,\delta_1)$ such that for each x in R^{k+1}, $F(x)|C_2 = f(x)$.

Suppose that f is a mapping of S^{k+1} into $S_2(i_2,\delta)$. The sphere S^{k+1} is the union of two $(k + 1)$-cells R_1^{k+1} and R_2^{k+1} whose common part is a k-sphere S^k. Then $f_1 = f|R_1^{k+1}$ and $f_2 = f|R_2^{k+1}$ are mappings in $S_2(i_2,\delta)$; thus there are mappings F_1 and F_2 of R_1^{k+1} and R_2^{k+1} into $S_1(i_1,\delta_1)$ such that $F_1(x)|C_2 = f_1(x)$ for x in R_1^{k+1} and $F_2(x)|C_2 = f_2(x)$ for x in R_2^{k+1}.

Consider $S^{k+1} \times C_1$. Denote by f^* the homeomorphism of $S^{k+1} \times (C_2 \cup$ bdry $C_1)$ into $S^{k+1} \times C_1$ which is such that $f^*(x,y) = (x,f(x)(y))$ for y in C_2 and $f^*(x,y) = (x,y)$ for y in bdry C_1 and by F_i^*, $i = 1,2$, the homeomorphism of $R_i^{k+1} \times C_1$ into itself which is such that $F_i^*(x,y) = (x,F_i(x)(y))$ for (x,y) in $R_i^{k+1} \times C_1$. Then $d[(x,y),\ f^*(x,y)] < \delta$, $d[(x,y),F_i^*(x,y)] < \delta_1$ and F_i^* extends $f^*|R_i^{k+1} \times (C_2 \cup$ bdry $C_1)$. The mappings F_i^* and F_2^* may not agree on $S^k \times C_1$. If they do not, consider a k-sphere S_2^k in int R_2^{k+1} bounding a $(k + 1)$-cell T_2^{k+1} such that $cl(R_2^{k+1} - T_2^{k+1})$ is homeomorphic to the product $S^k \times I$. The argument used in the proof of part (1) may be applied to construct an ϵ-homeomorphism F^* of $S^{k+1} \times C_1$ onto itself which is such that F^* extends f^*, $F^*|R_1^{k+1} \times C_1 = F_1^*$ and $F^*|T_2^{k+1} \times C_1 = F_2^*|T_2^{k+1} \times C_1$. Then the mapping F of S^{k+1} into H_1 which is such that $F^*(x,y) = (x, F(x)(y))$ is continuous and is such that for each x in S^{k+1}, $F(x)$ is an ϵ-homeomorphism which extends $f(x)$. Thus part (2) of Theorem 5.1 is true for $n = k + 1$.

$\underline{\text{Proof of Theorem 5.2 for } n = k + 1.}$ The proof follows rather closely that of Theorem 1 of [6] and that of Lemma 5.4 of [7]. Let G be a cellular decomposition of M in the sense of Definition 4.20. The proof will follow by induction on the number of cells in G. If G consists of only one cell, then M is a 3-cell and the theorem follows from the above mentioned theorem of Alexander.

Suppose that Theorem 5.2 is true for $n = k + 1$ and all compact 3-manifolds with boundary which have a cellular decomposition with fewer than r elements and that G is a cellular decomposition of M with r elements. Let C_1 be an element of G and let C_2 be a 3-cell in C_1 with the property that $\text{cl}(M - C_2)$ is homeomorphic to $\text{cl}(M - C_1)$ and $C_2 \cap \text{bdry } C_1 \subset \text{int } (C_1 \cap \text{bdry } M)$ and is homeomorphic to $C_1 \cap \text{bdry } M$. Then C_1 and C_2 satisfy the hypotheses of Theorem 5.1 which has been proven to be true for $n = k + 1$. Denote by H^*, H_1^*, and H_2^* the spaces of homeomorphisms of M, C_1, and $\text{cl}(M - C_2)$ onto themselves leaving their boundaries pointwise fixed, by i^*, i_1^* and i_2^* their identities, and by $S^*(i^*, \epsilon)$, $S_1^*(i_1^*, \epsilon)$, and $S_2^*(i_2^*, \epsilon)$ the (open) ϵ-neighborhoods of the identities.

If ϵ is a positive number, it follows from the induction hypothesis that there is a positive number δ' such that (1) each mapping of S^j, $j \leq k + 1$ into $S_1^*(i_1^*, \delta')$ is homotopic in $S_1^*(i_1^*, \epsilon/2)$ to the map of S^j into i_1^* and (2) each mapping of S^j, $j \leq k + 1$ into $S_2^*(i_2^*, \delta')$ is homotopic in $S_2^*(i_2^*, \epsilon/2)$ to the map of S^j into i_2^*. Let $\delta < \delta'$ be a positive number such that if f is a mapping of S^j, $j \leq k + 1$, into $S^*(i^*, \delta)$, then there is a mapping f^* of S^j into $S^*(i^*, \delta'/2)$ such that $f^*(x)|M - C_1 = i^*|M - C_1$ and $f^*(x)|C_2 = f(x)|C_2$. Clearly $f^*(x)|C_1$ is an element of $S_1^*(i_1^*, \delta'/2)$; hence there is a mapping F of $S^j \times I$ into $S^*(i^*, \epsilon/2)$ such that $F(x,1) = f^*(x)$, $F(x,0) = i^*$ and $F(x,t)|M - C_1 = i^*|M - C_1$ for each t. The homeomorphism $[f^*(x)]^{-1}f(x)|C_2 = i^*|C_2$; consequently $[f^*(x)]^{-1}f(x)|\text{cl}(M - C_2)$ is an element of $S_2^*(i_2^*, \delta')$. Thus there is a mapping F^* of $S^j \times I$ into $S^*(i^*, \epsilon/2)$ such that $F^*(x,1) = [f^*(x)]^{-1}f(x)$, $F^*(x,0) = i^*$ and $F^*(x,t)|C_2 = i^*|C_2$ for each t. Let $Z(x,t)$

denote $F(x,t)F^*(x,t)$. Then $Z(x,t)$ is an element of $S^*(i^*,\epsilon)$, $Z(x,1) = F(x,1)F^*(x,1) = f(x)$ and $Z(x,0) = i^*$. Thus H is LC^{k+1}, the induction is complete and Theorems 5.1 and 5.2 are proved.

<u>Corollary 5.7.</u> The space of homeomorphisms of a compact 3-manifold with boundary onto itself is LC^n for each n.

<u>Proof.</u> Let H denote the space of homeomorphisms of M onto itself, i its identity, and $S(i,\epsilon)$ the ϵ-neighborhood of the identity. Since H may be given a group structure, it need only be proved that H is LC^n at i. Let K denote the boundary of M, H* the space of homeomorphisms of K onto itself, i* its identity and $S^*(i^*,\epsilon)$ the ϵ-neighborhood of the identity. Let $H(M,K)$ denote the space of homeomorphisms of M onto itself leaving K pointwise fixed. It follows from Theorem 1 of [6] that H* is LC^n for each n.

There is a positive number $\delta' < \epsilon/2$ such that each mapping of S^j, $j \leqq n$, into $S(i,\delta') \cap H(M,K)$ is homotopic in $S(i,\epsilon/2) \cap H(M,K)$ to the mapping of S^j onto i and there is a positive number $\delta < \delta'/2$ such that each mapping of S^j, $j \leqq n$, into $S^*(i^*,\delta)$ is homotopic in $S^*(i^*,\delta'/6)$ to the mapping of S^j onto i*. Let z be a homeomorphism of $K \times I$ into M such that $z(x,1) = x$ for each x and $d(x,z(x,t)) < \delta'/6$ for each (x,t).

Suppose that f is a mapping of S^j, $j \leqq n$, into $S(i,\delta)$. There is a mapping F* of $S^j \times I$ into $S^*(i^*, \delta'/6)$ such that $F^*(x,0) = i^*$ and $F^*(x,1) = f(x)|K$. For each (x,t) in $S^j \times I$ and y in M, define $p(x,t)(y)$ so that (1) if $y \in M - z(K \times I)$, then $p(x,t)(y) = y$ and (2) if $y = z(a,s)$, then $p(x,t)(y) = z[F^*(x,st)(a),s]$. It is readily seen that $p(x,t) \in S(i, \delta'/2)$, that $p(x,t)|K = F^*(x,t)$, and that $p(x,p) = i$. Consider $[p(x,1)]^{-1}f(x)$. This is an element of $H(M,K) \cap S(i,\delta')$; thus there is a mapping g of $S^j \times I$ into $H(M,K) \cap S(i,\epsilon/2)$ such that $g(x,0) = i$ and $g(x,1) = [p(x,1)]^{-1}f(x)$. Let $F(x,t)$ be a mapping of $S^j \times I$ into $S(i,\epsilon)$ defined so that $F(x,t) = p(x,2t)$, $0 \leqq t \leqq 1/2$, and $F(x,t) = p(x,1)g(x,2t - 1)$ for $1/2 \leqq t \leqq 1$. Then $F(x,0) = i$, $F(x,1 = f(x)$ and the proof that H is LC^n is complete.

6. <u>Some consequences of the preceding theory.</u>

<u>Theorem 6.1.</u> Suppose that f is an h - 2-regular mapping of a complete metric space X onto a finite (covering) dimensional space Y such that each inverse under f is homeomorphic to the compact 3-manifold with boundary M and that either the Poincaré conjecture is true or each homotopy cell in M is a 3-cell. Then (X,f,Y) is a locally

trivial fibre space. If Y is locally compact, separable and contractible, then X is homeomorphic to $Y \times M$, where f corresponds to the projection map of $Y \times M$ onto Y.

Proof. It follows from Theorem 4.24 that f is completely regular. Let $G(y)$ denote the space of homeomorphisms of M into $f^{-1}(y)$ and Q the mapping of $\bigcup G(y)$ onto Y carrying $G(y)$ onto y. It follows from Theorem 5.2 that $G(y)$ is LC^n for each n and thus that the collection of all $G(y)$ is equi-LC^n (see section 5 above and Lemma 3 of [5]). Also, $\bigcup G(y)$ may be metrized so as to be complete and the collection of all $G(y)$ is lower semi-continuous (Q is open). Thus Theorem M may be applied to prove that if $y \in Y$, then there is an open neighborhood U_y of y and a mapping q_y of U_y into $\bigcup G(y)$ such that if $y' \in U_y$, then $q_y(y') \in G(y')$. Let q_y^* denote the homeomorphism of $U_y \times M$ onto $Q^{-1}(U_y)$ such that $q_y^* (y',x) = (y', q_y(y')(x))$. Hence (X,f,Y) is a locally trivial fibre space. That X is homeomorphic to $Y \times M$ if Y is locally compact, separable and contractible, follows from [17], p. 53.

Theorem 6.2. If, under the conditions of Theorem 6.1, f_y denotes a homeomorphism of M onto $f^{-1}(y)$ and K denotes bdry M, then if there is a homeomorphism h of $\bigcup f_y(K)$, $y \in Y$, onto $Y \times K$ such that the diagram

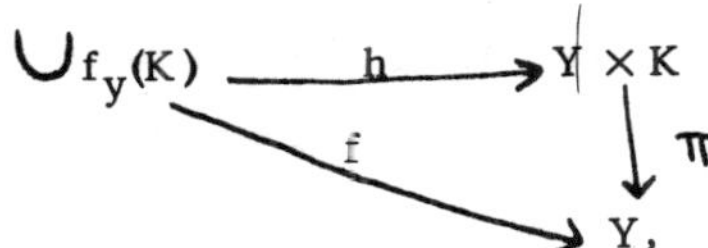

where π is the projection map, is commutative, then each point of Y is contained in a neighborhood U for which there is a mapping h^* of $f^{-1}(U)$ onto $U \times M$ that extends $h|\bigcup f_y(K)$, $y \in U$, such that the diagram

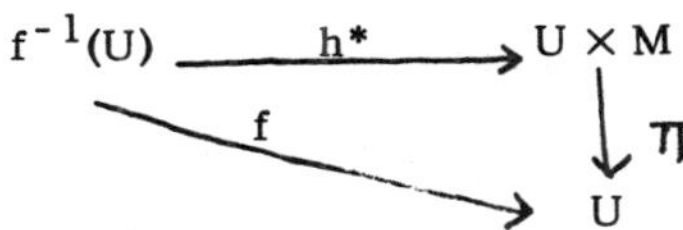

is commutative.

Proof. It follows as in the proof of Lemma 5.3 that f is completely regular in the strong sense that if $y \in Y$ and $e > 0$, then there is a $\delta > 0$ such that if $d(y,y') < \delta$, then there is a homeomorphism z of $f^{-1}(y)$ onto $f^{-1}(y')$ such that z moves no point as much as e and $z f_y |K = f_{y'}|K$. Also, if $G(y)$ is the space of homeomorphisms of M onto

$f^{-1}(y)$ which extend $f_y|K$, then the collection of all $G(y)$ is equi-LC^n and lower semi-continuous and $\bigcup G(y)$ may be metrized so as to be complete. Thus, as in the proof of Theorem 6.1, Theorem M may be applied to obtain Theorem 6.2.

Corollary 6.3. If (1) under the hypotheses of Theorem 6.1, the space of homeomorphisms of M onto itself has the property that its homotopy groups of order $\leq n$ all vanish or (2) under the hypotheses of Theorem 6.2, the space of homeomorphisms of M onto itself leaving K pointwise fixed has the property that its homotopy groups of order $\leq n$ all vanish, then if the covering dimension of Y does not exceed $n + 1$, the set U in the conclusions of 6.1 and 6.2 may be taken to be the entire space Y.

Proof. This is a consequence of the last sentence in the statement of Theorem M.

Corollary 6.4. Under the hypotheses of Theorem 6.1, if Z is a closed subset of Y and there is a homeomorphism h_1 of $f^{-1}(Z)$ onto $Z \times M$ such that the diagram

$$
\begin{array}{ccc}
f^{-1}(Z) & \xrightarrow{\ h_1\ } & Z \times M \\
& \searrow{\scriptstyle f} & \ \downarrow{\scriptstyle \pi} \\
& & Z
\end{array}
$$

is commutative, then there is an open set U in Y which contains Z for which there is a homeomorphism h^* of $f^{-1}(U)$ onto $U \times M$ which extends h_1 and is such that the diagram

$$
\begin{array}{ccc}
f^{-1}(U) & \xrightarrow{\ h^*\ } & U \times M \\
& \searrow{\scriptstyle f} & \ \downarrow{\scriptstyle \pi} \\
& & U
\end{array}
$$

is commutative.

Corollary 6.5. Under the hypotheses of Theorem 6.2, if Z is a closed subset of Y and h_1 is a homeomorphism of $f^{-1}(Z)$ onto $Z \times M$ such that the diagram

$$
\begin{array}{ccc}
f^{-1}(Z) & \xrightarrow{\ h_1\ } & Z \times M \\
& \searrow{\scriptstyle f} & \ \downarrow{\scriptstyle \pi} \\
& & Z
\end{array}
$$

is commutative and h_1 extends $h|\bigcup f_z(K)$, $z \in Z$, then U and h^* may be so chosen that h^* extends $h|\bigcup f_u(K)$, $u \in U$.

REFERENCES

1. J. W. Alexander, "On the deformation of an n-cell", Proceedings of the National Academy of Sciences, U. S. A., vol. 9 (1923), pp. 406-407.

2. Paul Alexandroff and Heinz Hopf, "Topologie", Berlin, 1935.

3. R. H. Bing, "A characterization of 3-space by partitionings", Transactions of the American Mathematical Society, vol. 70 (1951), pp. 15-27.

4. G. M. Fisher, "On the group of all homeomorphisms of a manifold", Transactions of the American Mathematical Society, vol. 97 (1960), pp. 193-212.

5. M.-E. Hamstrom and Eldon Dyer, "Completely regular mappings", Fundamenta Mathematicae, vol. 45 (1958), pp. 103-118.

6. _______________________________, "Regular mappings and the space of homeomorphisms on a 2-manifold", Duke Mathematical Journal, vol. 25 (1958), pp. 521-532.

7. M.-E. Hamstrom, "Regular mappings whose inverses are 3-cells", American Journal of Mathematics, vol. 82 (1960), pp. 393-429.

8. J. M. Kister, "Isotopies on 3-manifolds with boundaries, I and II", Transactions of the American Mathematical Society vol. 97 (1960), pp. 213-224.

9. V. L. Klee, Jr., "Some topological properties of convex sets", Transactions of the American Mathematical Society, vol. 78 (1955), pp. 30-45.

10. E. A. Michael, "Continuous selections, II", Annals of mathematics, vol. 64 (1956), 562-580.

11. Edwin E. Moise, "Affine structures in 3-manifolds, III, Tubular neighborhoods of linear graphs", Annals of Mathematics, vol. 55 (1952), pp. 203-214.

12. _______________, "Affine structures in 3-manifolds, V, The triangulation theorem and hauptvermutung", Annals of Mathematics, vol. 56 (1952), pp. 96-114.

13. C. D. Papakyriakopoulos, "On Dehn's Lemma and the asphericity of knots", Annals of Mathematics, vol. 66 (1957), pp. 1-26.

14. J. H. Roberts, "Local arcwise connectivity in the space H_n of homeomorphisms of S^n onto itself", Summary of Lectures, Summer Institute on Set Theoretic Topology, Madison, Wisconsin, 1955, p. 100.

15. D. E. Sanderson, "Isotopy in 3-manifolds, III, Connectivity of spaces of homeomorphisms", Proceedings of the American Mathematical Society, vol. 11(1960), pp. 171-176.

16. H. Seifert and W. Threlfall, "Lehrbuch der Topologie", Leipzig, 1934.

17. Norman Steenrod, "The topology of fibre bundles", Princeton, New Jersey, 1951.

18. Arnold Shapiro and J. H. C. Whitehead, "A proof and extension of Dehn's Lemma", Bulletin of the American Mathematical Society, vol. 64 (1958), pp. 174-178.

19. J. H. C. Whitehead, "Simplicial spaces, nuclei and m-groups", Proceedings of the London Mathematical Society, vol. 45 (1939), pp. 243-327.

20.   J. H. C. Whitehead, "On 2-spheres in 3-manifolds", Bulletin of the American Mathematical Society, vol. 64 (1958), pp. 161-166.

21.   G. T. Whyburn, "On sequences and limiting sets", Fundamenta Mathematicae, vol. 25 (1935), pp. 408-426.

Goucher College